Heat Transfer Phenomena and Applications

Heat Transfer Phenomena and Applications

Editor

Sunan Metharom

scitus
academics

Heat Transfer Phenomena and Applications
Edited by **Sunan Metharom**

ISBN: 978-1-68117-518-8
Library of Congress Control Number: 2016942878

SCITUS Academics LLC,
www.scitusacademics.com
Box No. 4766, 616 Corporate Way,
Suite 2, Valley Cottage,
NY 10989

Notice

Printed in United States of America on Acid Free Paper

Preface

Heat transfer is one of the three key supporters of chemical engineering and engineering science and contains many simple and practical principles that are used in countless industries. Heat transfer deals with transfer of energy in the form of heat; applications occur almost exclusively with heat exchangers used in chemical, petrochemical, petroleum refinery and engineering processes. The heat transfer takes place between a warm and a cold body, which is usually called the source and the receiver. Heat is associated with a system's internal potential and kinetic energy. Heat transfer or possibly dispersion can occur through three main mechanisms, conduction, radiation and convection. Heat moves from a higher temperature region to areas with lower temperature during conduction. This happens in solid, fluid or possibly gaseous mediums or between different mediums that make direct physical contact with each other. In convection, the combined effect of heat conduction, energy storage and mixing movement contributes to the transport of energy. Heat moves from a higher temperature body to a lower temperature body when the bodies are separated in space, even though a vacuum.

This interdisciplinary book covers issues related to combined heat transfer action and concurrent processes. This book also touches on the methodical solution for heat transfer problems. In this subject, the study of heat transfer phenomena and applications is equally emphasized. The text would hopefully be a valuable tool for people involved in heat transfer applications directly or indirectly in industry and academia.

Table of Contents

1

CHAPTER

Methodology to Calculate Boundary Conditions in a Single Isolated Helically Segmented Finned Tube Module

E. Martínez[1], G. Soto[1], W. Vicente[2], M. Salinas[2] and A. Campo[3]

[1] *Universidad Autónoma Metropolitana – Azcapotzalco, Mexico City, Mexico*

[2] *Instituto de Ingeniería, Universidad Nacional Autónoma de México, Ciudad Universitaria, Mexico City, Mexico*

[3] *Department of Mechanical Engineering, University of Texas at San Antonio, San Antonio, TX, USA*

1. INTRODUCTION

Helically segmented finned tubes are used in compact heat recoveries in order to save energy in industrial applications. These equipments are small because the gas phase turbulence and the heat transfer surface are increased by the presence of fins; both are relevant in heat transfer. However, the gas phase pressure drop is elevated and consequently, operational problems such as backpressure can emerge. Therefore, a study focusing on finned tubes is important in order to understand the fluid dynamics and heat transfer phenomena. There are two main methods for the analysis; the first uses integral analysis (gross effects) and the second uses Computational Fluid Dynamics (CFD) techniques. Integral analyses allow a quick evaluation of thermo-physical phenomena with minimum computational infrastructure but only gross effects can be examined. These analyses are primarily used in the design of equipment because only inlet and outlet fluid properties are important. The CFD technique requires good computational support and long calculation times, but it provides complete and detailed information on the intricate thermo-physical phenomena. This modern analysis requires a correct implementation of boundary conditions in order to adequately represent the flow hydrodynamics and heat transfer phenomena.

The implementation of boundary conditions is relevant in differential analyses because predictions depend from it. The differential analyses can be developed by means of analytical solutions or by means of numerical methods. In the case of helically segmented finned tube bank analyses, the analytical solutions are not possible because the geometry is complex. Then, a numerical simulation of helically segmented finned tube bank is the best option. The numerical simulations can be carried out by means of three different CFD alternatives such as Direct Numerical Simulation (DNS), Large Eddy Simulation (LES), and Reynolds Average Navier-Stokes Equations (RANS). The DNS technique is limited to low Reynolds flows with simple configurations. The LES technique is less demanding than the DNS, but it takes considerable computing resources and computing time because the required calculations are always three-dimensional and unsteady. Finally, the RANS technique, which is widely used in industrial applications, considers average spatial and temporal scales of turbulent fluctuations and solves the transport equations as a function of these average variables. However, these equations are not closed and additional models (turbulence models) are indispensable to close the system. So, the numerical analysis on helically segmented finned tube bank (complex geometry) is proposed with the RANS technique.

In the open literature there are several papers have been focused on numerical analysis of small finned tube banks, the majority of them are restricted to numerical simulations on bare tube layout (symmetric tube layout). For example, Beale and Spalding [1], Comini and Croce [2] and Beale [3, 4] have performed simulations on laminar flow regimes exclusively. Other authors like Benhamadouche and Laurennce [5] and Salinas-Vazquez et al. [6] have conducted studies on turbulent flow regimes. These simulations have analyzed symmetric tube layout with periodic boundary conditions. On the other hand, there are few numerical simulations focused on helically segmented finned tubes (asymmetric finned tubes). The papers are focused on laminar flows with Dirichlet boundary conditions. For example, Hofmann [7] and Mcilwain [8, 9] conducted two-dimensional simulations on a single helically segmented finned tube. Afterwards, Lemouedda et al. [10] developed a three-dimensional numerical simulation in a small finned tube bank. There are no reports of numerical simulations of asymmetric finned tube layout (helically segmented finned tube bank) under periodic boundary conditions and the effect of inside fluid temperature has not been considered. Therefore, a methodology to calculate flow properties in different zones of finned tube bank is required in order to implement boundary conditions on a single isolated finned tube module. This methodology considers calculations in entire and partial finned tube layout and it is applied to calculate boundary conditions in a numerical simulation. Then, a compact heat recovery in staggered layout is represented as some single isolated finned tube modules in order to save computational resources. The single isolated finned tube module is simulated and predictions are compared with results from correlations available in the open literature.

2. METHODOLOGY

The differential analysis of compact heat recoveries with CFD techniques requires high calculation times because a full finned tube bank needs to be simulated. The dimensions of computational domain are high due to size of finned tubes and number of finned tubes used in the equipment. So, the necessity to reduce calculation times is relevant in numerical simulations because these times can be excessively-high. The only way to reduce calculation times is by reduction of computational domain but it requires a correct implementation of boundary conditions for representing adequately physical phenomena. A complete finned tube bank can be represented by a single isolated finned tube module in the fully developed flow region as shown in Figure 1. The single isolated finned tube module consists of an arrangement of entire and partial finned tubes. This finned tube layout can allow a computational domain reduction of 99% but requires values of boundary conditions in intermediate regions of finned tube bank. The values of boundary conditions for velocity, pressure, and temperature should be calculated for entire and partial finned tubes in intermediate regions of finned tube bank. Then, a method to calculate those boundary conditions must be developed in order to represent the finned tube bank as a single isolated finned tube module (Figure 1). This methodology is based on integral models which have been validated experimentally [22] with precision higher than 90% for pressure and 95% for temperature. So, a numerical simulation of single isolated finned tube module in the fully developed flow with periodic boundary conditions is done. Numerical predictions are compared with results obtained from best correlations available in the open literature, which are presented in section 2.2. So, the mean pressure drop, mean temperature difference, mean Nusselt number and mean friction factor are compared.

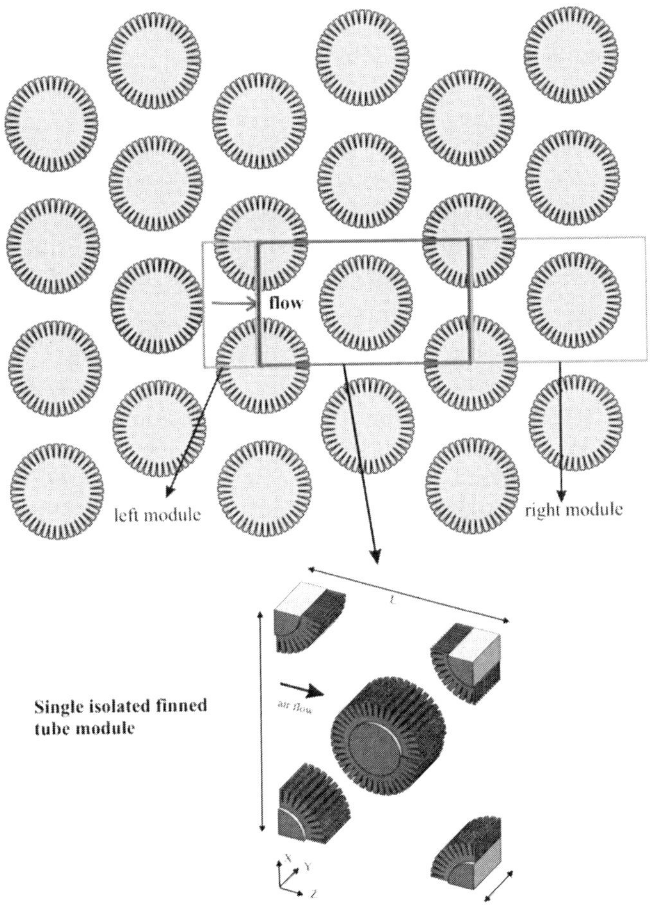

Figure 1. Finned tube bank in staggered layout.

2.1. Finned Tube Bank Analysis

The complete finned tube bank may be represented as a single isolated fined tube module in the fully developed flow as shown in Figure 1. This finned tube layout requires values for velocity, pressure, and temperature in intermediate regions of finned tube layout for both outside and inside flows. These values are obtained by means of an integral analysis, which considers the Logarithmic Mean Temperature Difference method (LMTD). However, this method only can be applied for entire finned tubes in arrangement of two or more finned tube rows and two or more finned tubes per row. Then, the complete finned tube bank and the single isolated finned tube module should be analyzed in order to apply LMTD method, which is described in section 2.2. The finned tube bank can be divided in single isolated finned tube modules as shown in Figure 2a. In this figure , a full finned tube bank (6 finned tube rows with 4 finned tubes per row) is composed by 12 single isolated finned tube modules (black and red boxes). Every finned tube module contains 2 finned tubes because there are one entire finned tube and 4 quarters of finned tube. Then, two single isolated finned tube modules like (red boxes) can be represented as an equivalent small finned tube bank (blue box) as shown in Figure 2b. Therefore, the finned tube bank showed in Figure 1 is represented as a single isolated finned tube modules arrangement as presented in Figure 2a.

Figure 2b shows a small finned tube bank (blue box) composed of 2 finned tubes per row and 2 rows of finned tubes, which is arranged in order to obtain 2 single isolated finned tube modules (red box). Then, the analysis of single isolated finned tube modules must consider minimum arrangements of 2 finned tubes per row and 2 rows of finned tubes. This consideration does not affect predictions of friction factor and Nusselt number because these dimensionless parameters are not function of number of finned tubes involved in the arrangement if mass flow is corrected to the new finned tube layout, which is demonstrated in the sensitive analysis (section 4.1). So, models for evaluating heat transfer and pressure drop can be applied to the equivalent small finned tube bank (Figure 2b). The pressure drop depends mainly of flow hydrodynamics, which shows similar velocity fields for every single isolated finned tube module as is discussed in results analysis (section 4.2). Therefore, the models for evaluating pressure drop can be applied directly in the equivalence small finned tube bank (Figure 2b). However, the pressure drop cannot be considered as a boundary condition because only represents the pressure difference at the inlet and exit of single isolated finned tube modules. The pressure in the boundaries of single isolated finned tube module is obtained with the analysis of finned tube bank fromFigure 1. In this figure , the single isolated finned tube in the fully developed region (red box) is located near to the exit of module. This finned tube module has to the right a part of finned tube module (blue box) while at left has one and a part finned tube module (green box). In the case of an atmospheric discharge of flow gases (zero relative pressure), which is correct because flue gases of compact heat recoveries cannot be used in additional industrial process, the relative pressure drop in the last part of single isolated finned tube module (blue box, Figure 1) is calculated as an proportional arithmetic mean pressure drop to the part of this finned tube module. The proportional part of this module corresponds to a value of 0.75. Therefore, the relative pressure at the exit of the single isolated finned tube module in the fully developed flow (red box, Figure 1) corresponds to the value calculated previously. Finally, the relative pressure at the inlet of the single isolated finned module (red box, Figure 1) is calculated from the sum of the pressure drop in this finned tube module and the proportional part (0.25) of the left finned tube module (green box, Figure 1).

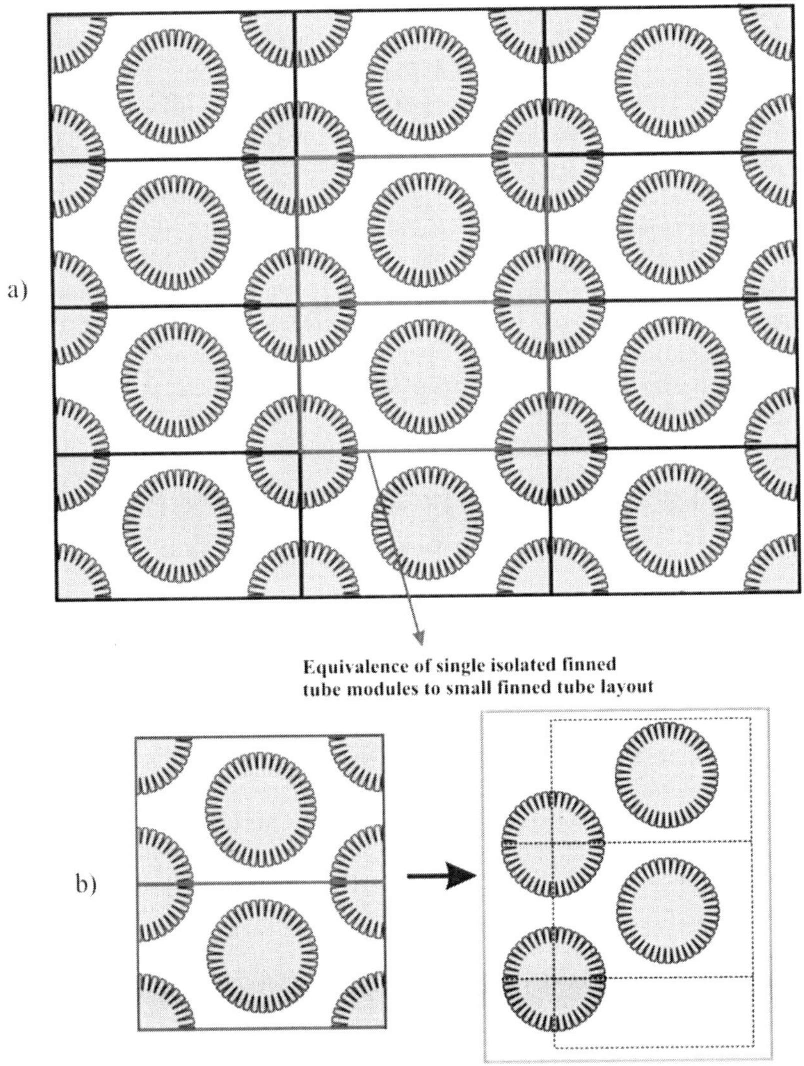

Figure 2. Single isolated finned tube modules in the finned tube bank.

The direct application of LMDT method for heat transfer in the equivalent small finned tube bank (blue box, Figure 2b) is not recommended because the outside flow, in the single isolated finned tube module (red box, Figure 2b), is cooled by tree different cooling sources (inside fluid temperature). While the equivalent small finned tube bank (blue box, Figure 2) only has the influence of two cooling sources. One way to solve this problem is by means of temperature evaluation of outside flow for small finned tube layouts. The finned tube layouts considered have an initial arrangement of two finned tube rows, which are evaluated. Later, calculations over initial finned tube bank with additional finned tube rows are proposed (see section 2.3). This procedure is iterative because only initial conditions at the inlet of gas-phase flow and inside fluid in the complete finned tube bank (Figure 1) are known. Once temperature evaluation for each finned tube layout is done, the inside fluid temperatures for each finned tube row are defined. The sensitivity analysis (section 4.1) of a finned tube bank shows that consideration applied for evaluating boundary conditions are satisfactory. These boundary conditions are employed in a numerical analysis of the finned tube bank showed in Figure 1, which is described in section 2.3. Numerical predictions are compared with results from models for heat transfer and pressure drop in helically segmented finned tubes presented in section 2.2. The comparative analysis show close values between models and numerical results as discussed in results (section 4).

2.2. Logarithmic Mean Temperature Difference (Lmtd) Method

The methodology for calculating boundary conditions in the single isolated finned tube module is based on the Logarithmic Mean Temperature Difference method. This method allows the evaluation of heat transfer coefficients and friction factors for different geometries, according to models available in the open literature. These parameters permit the global evaluation of heat transfer and flow hydrodynamics of finned tube bank. The LMTD method considers the evaluation of overall heat transfer coefficient (U), which is based on the outside finned tube and is defined with the following equation:

$$U_o = \frac{1}{\dfrac{A_o + A_o R_{fo}(h_o + h_r)}{(h_o + h_r)(\eta_f A_f + A_t)} + \dfrac{e_w A_o}{k_w A_i} + \left(\dfrac{1}{h_i} + R_{fi}\right)\dfrac{A_o}{A_i}} \quad (1)$$

where *ho*, *hi*, and *hr* are mean outside convective coefficient, mean inside convective coefficient, and radiation heat transfer coefficient, respectively. In the case of flue gas temperatures lower than 300ºC, the value of *hr* could be negligible [11], and so this value is considered zero. *Rfo* and *Rfi*, are the outside and inside fouling factors, respectively. *ηf*, *Af*, *At*, *Ao*, and A_i are fin efficiency, fin surface area, bare tube surface area, total surface area, and inside surface area, respectively. Finally, *ew* and *kw* are tube wall thickness and tube material thermal conductivity, respectively.

The mean convective coefficients are calculated for the inside and outside of finned tubes. The mean inside convective coefficient (*hi*) considered in the evaluation of *U* is the Gnielinski's correlation [12]. This model has been validated with satisfactory results by Rane, et al. [13] and according to Bejan [14] is the best available in the open literature. The Gnielinski's correlation [12] is valid for $0.5 \le Pr \le 2000$ and $3000 \le Re \le 5x10^6$, which is shown in the following equation:

$$Nu = \frac{h_i d_i}{k} = \frac{(f_i / 8)(\mathrm{Re} - 1000)\mathrm{Pr}}{1 + 12.7(f / 8)^{1/2}(\mathrm{Pr}^{2/3} - 1)} \quad (2)$$

where *di* and *k* are inside diameter of tube and thermal conductivity of fluid. Re and Pr are the Reynolds Number and Prandtl Number. Finally, *fi* is the friction factor, which is defined in the following equation:

$$f_i = \frac{1}{(1.82 \log_{10} \mathrm{Re} - 1.64)^2} \quad (3)$$

There are many mean outside convective heat transfer coefficients (*ho*) in the open literature such as models proposed by Weierman [15], ESCOA [16], Nir [17], and Kawaguchi et al. [18]. The models attributable to Weierman [15] and Kawaguchi et al. [18] are recommended by Martinez et al. [19], but Kawaguchi´s et al. [18] model is adopted. The model of Kawaguchi, et al. [18] is valid for $7000 \le Rev \le 50000$ and $0.112 \le sf/dv \le 0.198$, which is shown in the next equation:

$$Nu = A_2 \mathrm{Re}_v^{0.784} \mathrm{Pr}^{1/3} \left(s_f / d_v\right)^{-0.062} = \frac{h_o d_v}{k_g} \quad (4)$$

where *Rev* is the Reynolds number based on the volume-equivalent diameter. The terms *A2*, *sf*, and *dv*are the experimental coefficients for tube rows, fin gap, and volume-equivalent diameter, respectively. The coefficient A_2 is obtained from Kawaguchi´s, et al. [18] model. Finally, k_g is the thermal conductivity of gases.

The volume-equivalent diameter, *dv*, is defined by the following equation:

$$d_v = \left[t_f n_f \left\{ \left(d_o + 2l_f \right)^2 - d_o^2 \right\} + d_o^2 \right]^{1/2} \tag{5}$$

where *nf, tf, lf,* and *do* are fin number per unit length, fin thickness, fin height, and outside diameter of bare tube, respectively.

Gas phase pressure drop can be calculated with models proposed by Weierman [15], ESCOA [16], Nir [17], and Kawaguchi et al. [18]. The model of Weierman [15] has been validated with satisfactory results by Martinez, et al. [19] and as a consequence, is adopted. In the analysis of compact heat recoveries a maximum pressure drop of 248.9 Pa [20] (1 inch (in) of water column (wc)) is considered in order to avoid technical problems such as backpressure. The pressure drop is calculated with the following empirical equation:

$$\Delta P_g = \frac{\left(f_o + \frac{\left(1 + B^2\right)\rho_{gp}}{4N_r} \right) G_o^2 N_r}{1.083x10^9 \rho_{gp}} \tag{6}$$

where *B, Go, Nr, fo,* and *ρgp* are the contraction factor, the gas mass flux, the number of tube rows, the friction factor, and the gas-phase density at the average outside temperature, respectively.

The gas phase friction factor (*fo*) is calculated with Weierman´s model [15], which is valid for tube diameters between 38.1-60.96 mm and mass velocity of the gas between 0.67-40.36 kg/m^{2}s. The Weierman´s model [15] is shown in the following equation:

$$f_o = \left[0.07 + 8\,\mathrm{Re}_o^{-0.45} \right] 0.11 \left(\frac{0.05S_t}{d_o} \right)^{-0.7\left(\frac{l_f}{s_f}\right)^{0.23}}$$

$$\left[1.1 + \left(1.8 - 2.1e^{-0.15N_r^2}\right) e^{-\frac{2S_l}{S_t}} - \left(0.7 - 0.8e^{-0.15N_r^2}\right) e^{-\frac{0.6S_l}{S_t}} \right] \left(\frac{d_f}{d_o} \right)^{1/2} \tag{7}$$

where *Reo* is the Reynolds number based on the outside bare and *cp* is the specific heat capacity at constant pressure. Also, *df* and *do* are the outside diameter of the finned tube and the outside diameter of the bare tube, respectively. The terms *Sl* and *St* indicate the longitudinal pitch and the transverse pitch, respectively.

For more details about thermal evaluation of helically segmented finned tube banks see reference [21].

2.3. Sensitive Analysis Of Finned Tube Bank

The methodology is based on the equivalence of a finned tube bank with single isolated finned tubes modules layouts. This methodology, which is based on the LMTD method, requires that predictions of friction factor and Nusselt number be independent of finned tube layout and number of finned tubes involved in compact heat recoveries. Then, a sensitive analysis of the finned tube bank is required in order to validate considerations in the methodology. The sensitive analysis is done for a small heat recovery (Figure 1), in which results of friction factor and Nusselt number are compared for different finned tube layouts and finned tube lengths. The geometric characteristics of helically segmented finned tube and finned tube layout are shown in Figure 3. In this figure , a front and side views of the finned tube used in this study and its staggered configuration is presented. The finned tube bank analyzed corresponds to the layout shown in Figure 1, which is composed by 4 finned tubes per row and 6 finned tube rows. The small heat recovery (Figure 1) is analyzed with LMTD method for different finned tube configurations.

The finned tube configurations used in the sensitive analysis considers an initial finned tube layout of 2 finned tubes per row and 2 finned tubes rows (red polygon in Figure 4). Subsequently, one finned tube row is added to reach 6 finned tube rows (red dash boxes in Figure 4). Then, another finned tube is added to the row (blue polygon, Figure 4) in order to obtain 3 finned tubes per row and 2 finned tube rows. Previous procedures in initial finned tube layout are done until 6 finned tube rows (blue dash boxes,Figure 4) are reached. Finally, one last finned tube row is added (green polygon in Figure 4) in order to get 4 finned tubes per row and 2 finned tube rows. The same procedure for previous finned tube layouts are done until full finned tube bank (gray dashed box) is reached. Once finned tube bank layout analysis is done, a similar study is proposed for different finned tube length. The procedure is the same as previous analysis but at different finned tube length. The finned tube lengths proposed are 1m, 0.5 m, and 0.05194 m. The last finned tube length corresponds to the value used in the numerical simulation. The results (section 4.1) show variations lower than 1% for Nusselt number and 3.6% lower for friction factor. These results confirm that considerations employed in this methodology (boundary condition calculation) are appropriate. The analysis is obtained with the thermodynamic conditions presented intable 1 for different finned tubes per row and finned tubes length which, shows deviations on Reynolds number based in the outside diameter tube (Re_o) lower than 0.4%. The variations in Re_o are due to adjustment of mass flow at different finned tube configurations.

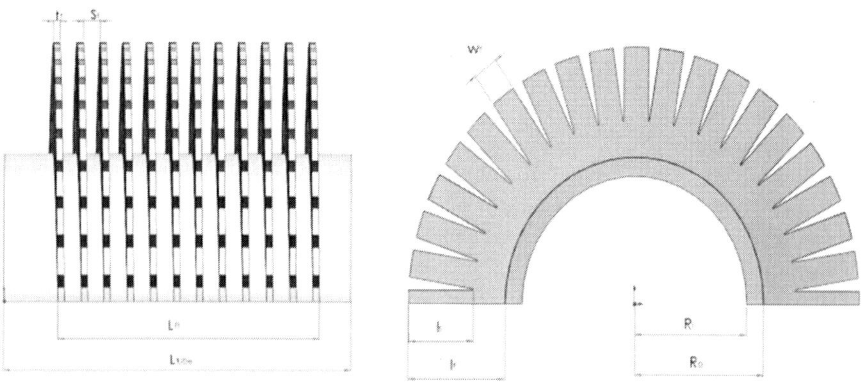

Tube		Fins		Finned tube layout	
D_o (m)	0.0508	l_f (m)	0.01905	S_T (m)	0.11430
D_i (m)	0.044	l_s (m)	0.01275	S_L (m)	0.09906
L_{tf} (m)	1.0	t_f (m)	0.0012	θ (°)	60
L_{th} (m)	0	s_f (m)	0.003	tube rows	6
L_{tube} (m)	1.0	w_f (m)	0.0048	tube per row	4
		d_f (fins/m)	236.0		

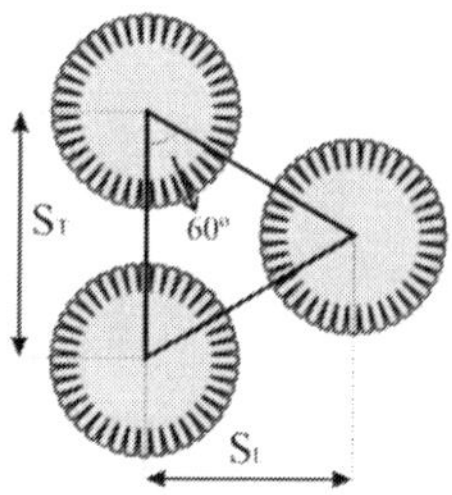

Figure 3. Geometric characteristics of finned tubes and finned tube layout.

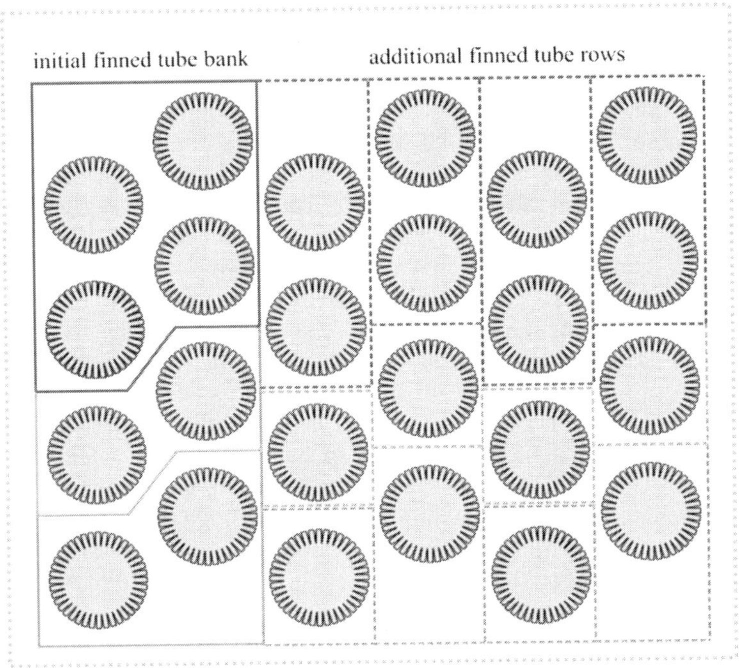

Figure 4. Finned tube layout configurations in sensitive analysis.

Table 1. Thermodynamic conditions.

Finned tube length: 1 m			
	Finned tubes per row		
Parameter	2	3	4
mass flow (kg/s)	0.3624	0.05436	0.7248
Re_o	8562	8562	8562
Ma	< 0.03	< 0.03	< 0.03
Temperature (ºC)	60	60	60
Finned tube length: 0.5 m			
	Finned tubes per row		
Parameter	2	3	4
mass flow (kg/s)	0.1812	0.2718	0.3624
Re_o	8567	8567	8567
Ma	< 0.03	< 0.03	< 0.03
Temperature (ºC)	60	60	60
Finned tube length: 0.05194 m			
	Finned tubes per row		
Parameter	2	3	4
mass flow (kg/s)	9.411×10^{-3}	14.116×10^{-3}	18.822×10^{-3}
Re_o	8599	8599	8599
Ma	< 0.03	< 0.03	< 0.03
Temperature (ºC)	60	60	60

3. NUMERICAL SIMULATION

In this section, the methodology for calculating boundary conditions in a single isolated finned tube module is employed in a numerical simulation of helically segmented finned tube bank in staggered layout. The objective is to apply the methodology in a numerical model that represents the interaction between the hydrodynamics and the heat transfer of a turbulent gas flow in complex systems. The simulation allows detailed analysis on compact heat recoveries that can be used for improving thermal behaviour. So, a correct implementation of boundary conditions in order to represent adequately physical phenomena is essential. The numerical model is focused to the outside finned tube, because the gas phase dominates the heat transfer [22] and also the pressure drop is critical [22]. However, the effect of the internal fluid is considered by means of an average inside temperature in each finned tube. Thus, the boundary conditions for outside flow and inside fluid are calculated with methodology described in section 2. In this numerical simulation, periodic boundary conditions are proposed for the single isolated finned tube module in the fully developed flow (Figure 1). Numerical predictions of Nusselt number and friction factor are compared with results obtained from correlations of Kawaguchi et al. [18] and Weierman [15], respectively in order to verify numerical results that depend on boundary conditions values.

3.1. Governing Equations

The analysis of turbulent flows is complex, because fluid properties are irregular in space and time. Instantaneous variables (φ) are represented as a function of two terms: the mean ($\tilde{\varphi}$) and its fluctuation (φ'')

These terms are used in the transport equations that govern the flow movement and the heat exchange. The transport equations are averaged by means of Favre [23] method, which are solved for mean values of the fluid properties. Thereby, the set of mass, momentum, and energy transport equations in Cartesian system is defined as follows:

$$\frac{\partial \bar{\rho}}{\partial t} + \nabla \cdot \left(\bar{\rho} \tilde{\vec{V}} \right) = 0 \tag{8}$$

$$\frac{\partial}{\partial t}\left(\bar{\rho} \tilde{\vec{V}} \right) + \nabla \cdot \left(\bar{\rho} \tilde{\vec{V}} \otimes \tilde{\vec{V}} \right) = -\nabla \cdot \bar{P} + \mu \nabla^2 \tilde{\vec{V}} - \nabla \cdot \left(\widetilde{\bar{\rho} \vec{V}'' \vec{V}''} \right) + \overline{\rho \vec{g}} \tag{9}$$

$$\frac{\partial}{\partial t}\left(\bar{\rho}\tilde{h}\right)+\nabla\cdot\left(\bar{\rho}\tilde{\vec{V}}\tilde{h}\right)=-\nabla\cdot\tilde{\vec{J}}_h-\nabla\cdot\left(\bar{\rho}\widetilde{\vec{V}''h''}\right) \qquad (10)$$

Where ρ¯, ˜V→˜, and ˜h˜ are the mean density, instantaneous velocity, fluctuating velocity, and mean enthalpy, respectively. μ and g→¯ are the viscosity and the gravitational acceleration, respectively. The terms

$\bar{\rho}\widetilde{\vec{V}''\vec{V}''}$ and $\bar{\rho}\widetilde{\vec{V}''h''}$ are the apparent Reynolds stress tensor and the turbulent heat flux, respectively. $\bar{\bar{P}}$ is the pressure tensor. Finally, $\nabla\cdot\tilde{\vec{J}}_h$ is the diffusive heat flux, which is modeled with Fourier's law. Equations (2) and (3) demand additional mathematical expressions to model $\widetilde{V''\varphi''}$ term. The closure of these equations requires modeling of Reynolds stress tensor and of the turbulent heat flux. Then, the Reynolds stress tensor is closed with turbulence models. In this work, the k-ε RNG (Renormalization Group) turbulence model developed by Yakhot and Orzag [24] is considered. The turbulent heat flux vector is obtained by means of an analogy between momentum transfer and thermal energy transfer. Under this concept, it is possible to establish a suitable articulation between the turbulent heat flux vector and the turbulent flow viscosity.

3.2. Boundary Conditions

The implementation of periodic boundary conditions alleviates the computational resources, because the computational domain is reduced considerably. So, only a single isolated finned tube module in the fully developed flow region needs to be simulated (Figure 1). Also, the tube length required (spatial direction y in Figure 1) is minimum because only is necessary the length for the flow does not vary in this spatial direction because periodic conditions assume no influence of walls or position in any direction. Then, the computational domain is reduced 99%, as shown in table 2, because only is necessary the single isolated finned tube module (Figure 1) with a tube length of 0.05194 m. However, the numerical simulation requires a correct inclusion of boundary conditions in order to adequately represent the physical phenomenon. So, the boundary conditions are applied to a turbulent air flow in a stationary and fully-developed flow regime. The methodology developed by Patankar et al. [25] and Kelkar and Patankar [26] is implemented in this work. This methodology was generated for laminar flows and as a consequence, only velocity, pressure and temperature are considered as a periodic behaviour. However, additional considerations for turbulent flows need to be implemented. These considerations should depend on the turbulence model selected.

The periodic velocity is based on the non slip condition on the boundary walls of the computational domain and negligible value of the velocity variation in every spatial direction. So, the analysis is presented for the flow direction (*z*), which can be generalized for the remaining spatial directions. Thereby, the periodic condition for the component velocity in the flow direction (w˜) is defined as follows:

$$\frac{\partial \tilde{w}}{\partial z} = 0 \qquad \tilde{u} = 0 \qquad \tilde{v} = 0 \tag{11}$$

where u˜,v˜,w˜ are the mean velocity components in the respective spatial directions *x*, *y*, *z*.

In the case of fully developed flows, the velocity is the same at a characteristic length (*L*), and this is shown in the next equation:

$$\tilde{w}(x,y,z) = \tilde{w}(x,y,z+L) \tag{12}$$

The periodic boundary velocities need an initial value *wb* (bulk velocity), which is calculated with the following equation:

$$w_b = \frac{1}{A_{yz}} \int_0^x \int_0^y w_{med}\, dx\, dy \tag{13}$$

where *wmed* is an initial velocity profile.

The pressure field is obtained from a periodic behavior taking into consideration a pressure drop in the flow direction. This implies that the pressure is defined by the sum of a periodic pressure term in the flow direction and an average pressure drop. This is represented in the following equation:

$$P(x,y,z) = \tilde{P}(x,y,z) - \beta z \tag{14}$$

where P˜ and β are periodic pressure and average pressure-gradient in the flow direction.

Table 2. Finned tube layout dimensions.

	Dimensions:			
Layout	x (m)	y (m)	z (m)	cells
Finned tube bank	0.4572	1.0	0.59878	34200000
Single isolated finned tube module	0.1143	0.05194	0.20312	178512
	Computational domain reduction:			
Finned tube bank	reference			
Single isolated finned tube module	99.6			

These two terms are defined in the pair of equations (15) and (16):

$$\tilde{P}(x,y,z) = \tilde{P}(x,y,z+L) \tag{15}$$

$$\beta = \frac{\tilde{P}(x,y,z) - \tilde{P}(x,y,z+L)}{L} = \frac{\Delta\tilde{P}}{L} \tag{16}$$

where ΔP˜ is the pressure drop over a finned tubes module, which can be calculated through the empirical equation (6).

The temperature field is obtained with constant wall heat flux boundary condition, which considers a constant variation of temperature in the flow direction; that is, the heat transfer magnitude is the same from one finned tube module to another finned tube module. This boundary condition can be appropriate for the present study because it can be applied to turbulent flows. However, it is appropriate if a uniform heat transfer is found in the small finned tube bank simulation. The numerical results in small finned tube bank show a quasi-constant mean temperature (plane xy) in the flow direction as shown in results section.

The temperature field is obtained from a periodic behavior and an adjustment term of the temperature in the flow direction. Thus, the temperature is defined by a periodic temperature term in the flow direction and an average temperature adjustment term expressed in the next expression:

$$T(x,y,z) = \tilde{T}(x,y,z+L) + \gamma z \qquad (17)$$

where T˜(x,y,z) and γ are the mean temperature field and the temperature-gradient term, respectively. The temperature-gradient term can be calculated with the following equation:

$$\gamma = \frac{\tilde{T}(x,y,z) - \tilde{T}(x,y,z+L)}{L} = \frac{\dot{Q}}{\dot{m}_g c_p L} \qquad (18)$$

Where Q˙,˙m˙, and *cp* are heat addition, mass flow, and specific heat at constant pressure, respectively.

The heat addition in the flow direction for the single isolated module is determined with the next equation:

$$\dot{Q} = U_o A_T \Delta T_{ML} \qquad (19)$$

where *Uo* is the overall heat transfer coefficient, *AT* the overall finned surface and *ΔTML* the logarithmic mean temperature difference.

The value of *Uo* is obtained from equation (1) and *AT* is obtained with geometry of finned tube. For more details about calculations of these parameters see reference [21].

The numerical simulation of turbulent flow on a single isolated module needs the implementation of periodic conditions for additional variables, according to the turbulence model. In the case of k-ε RNG turbulence model, the turbulent kinetic energy (*k*) and its dissipation rate (ε) show a periodic behavior, according to Martínez [27]. So, the periodic condition for the turbulent kinetic energy and its dissipation rate are defined as follows:

$$\tilde{k}(x,y,z) = \tilde{k}(x,y,z+L) \qquad (20)$$

$$\tilde{\varepsilon}(x,y,z) = \tilde{\varepsilon}(x,y,z+L) \tag{21}$$

The inclusion of periodic boundary conditions in the simulation of single isolated module requires light changes in the governing equations. The mass conservation equation does not change and is evaluated by means of equation (1). In the case of pressure field, the momentum conservation equation needs to include the adjust term of average pressure-gradient in the flow direction. So, the equation (2) can be written as:

$$\frac{\partial}{\partial t}\left(\bar{\rho}\tilde{\vec{V}}\right) + \nabla\cdot\left(\bar{\rho}\tilde{\vec{V}}\otimes\tilde{\vec{V}}\right) = \beta - \nabla\cdot\overline{P} + \mu\nabla^2\tilde{\vec{V}} - \nabla\cdot\left(\overline{\bar{\rho}\vec{V}''\vec{V}''}\right) + \overline{\rho\vec{g}} \tag{22}$$

The term β is included in the CFD code PHOENICS 3.5.1 [28] by means of an additional source term in the momentum equation. On the other hand, the evaluation of temperature field requires that energy conservation equation must be adjusted in order to include the cooling of gas phase in the flow direction. So, the equation (3) can be written as:

$$\frac{\partial}{\partial t}\left(\bar{\rho}\tilde{h}\right) + \nabla\cdot\left(\bar{\rho}\tilde{\vec{V}}\tilde{h}\right) + \nabla\cdot\left(\tilde{\vec{V}}\gamma\right) = -\nabla\cdot\tilde{\vec{J}}_h - \nabla\cdot\left(\overline{\bar{\rho}\vec{V}''h''}\right) \tag{23}$$

The term ∇·(V→˜γ) is included in the CFD code PHOENICS 3.5.1 [28] by means of an additional source term in the energy equation.

The turbulent kinetic energy and the dissipation turbulent rate are calculated directly from equations (20) and (21) because these equations do not change.

3.3. Numerical Details

The numerical simulation is developed for a single isolated finned tube module in the fully developed flow, which is shown in Figure 5. In this figure , the single isolated finned tube module is presented with boundary conditions calculated with methodology proposed. The dimensions of computational domain and the mesh used for simulation are shown in table 2. This table exhibits a reduction in computational domain with single isolated finned tube module of 99% that represents finite calculation times. The used mesh in numerical simulation considers that numerical predictions are independent from it. The thermodynamic employed conditions in the simulation work are presented in table 1 for finned tube length of 0.05194 m and Figure 5. On the other hand, the complex geometry is represented by cut-cell method [29], which allows the use of Cartesian grids. The numerical simulations consider a staggered grid under a hybrid discretization scheme of the convective term. It was also considered that the system is in a stationary state and is only exposed to one gravitational field (in heat recoveries, the fluid generally flows in vertical direction), and that the gas discharge (finned tube bank outlet) occurs in a sea level atmosphere. Finally, the geometric characteristics of the finned tube used in this study are shown concurrently in Figure 3.

The numerical results are used for evaluating average Nusselt Number (Nu) and average friction factor in small finned tube bank and single isolated module. The results are compared with values obtained from Correlations of Kawaguchi, et al [18] and Weierman [15] models, respectively.

The average Nusselt number is calculated with the next equation:

$$\overline{Nu} = \frac{\bar{h} d_v}{k_g} \tag{24}$$

where h¯ is the average convective coefficient, which is defined as:

$$\bar{h} = \frac{\dot{Q}}{A_o \Delta T_{LMnum}} \tag{25}$$

where *ΔTLMnum* is the numerical-logarithmic mean temperature difference, which is calculated as:

$$\Delta T_{LMnum} = \frac{\left[T_i(z+L) - T_b(z+L)\right] - \left[T_i(z) - T_b(z)\right]}{Ln \dfrac{\left[T_i(z+L) - T_b(z+L)\right]}{\left[T_i(z) - T_b(z)\right]}} \tag{26}$$

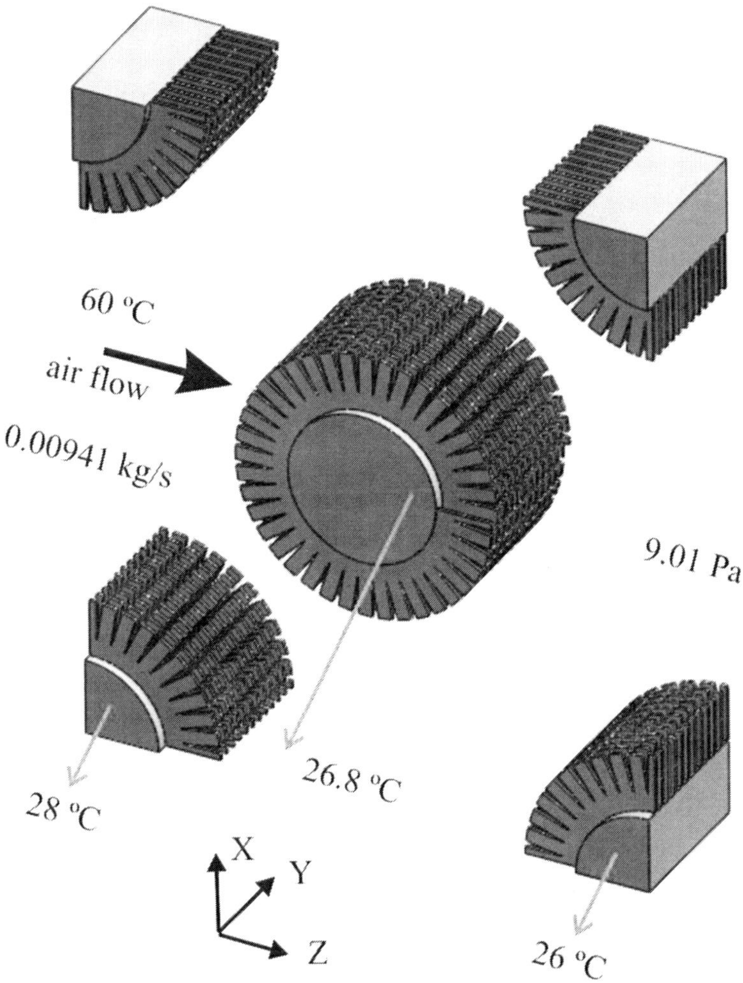

Figure 5. Computational domain in numerical simulation.

where *Ti* and *Tb* are the mean temperature inside finned tubes and mean boundary temperature, respectively. The conditions at *z* and *z+L* correspond to the inlet and exit of the single isolated finned tube module.

The numerical-average friction factor is obtained directly from equation (6) in which numerical pressure drop is calculated from the next expression:

$$\Delta\tilde{P} = P_b(z) - P_b(z + L) \qquad (27)$$

where Pb is the numerical-mean pressure.

4. RESULTS

The methodology proposed is applied for evaluating boundary conditions in a single isolated finned tube module (Figure 1). The predicted values are utilized in a numerical simulation in order to verify a correct representation of physical phenomena with these boundary conditions. These values require independence of results in heat transfer (Nusselt number) and pressure drop (friction factor) for different finned tube configurations, number of finned tubes, and finned tube lengths. So, a sensitive analysis is necessary in order to support previous considerations. The results of sensitive analysis shows that assumptions of methodology are correct, which are discussed in section 4.1. On the other hand, numerical predictions (focused in the outside flow) show a correct representation of interaction of heat transfer and flow hydrodynamics, which is presented in section 4.2. The comparative analysis of results between numerical predictions and results obtained from Kawaguchi et al [18] and Weierman [15] models for Nusselt number and friction factor, respectively, show close values as discussed in section 4.2.

4.1. Sensitive Analysis

The LMTD method is applied to different finned tube layouts and finned tube lengths, which are described in section 2.3. The main goal of the analysis is to verify that friction factor and Nusselt number are independent of finned tube rows, number of finned tube per row, and finned tube length for the finned tube bank showed in Figure 1. The comparative analysis at different finned tube configurations and finned tube lengths with thermodynamic conditions presented in table 1 are shown in table 4. In this

Table 3. Sensitivity analysis results.

Finned tube length: 1 m						
	Finned tubes per row					
	2		3		4	
Finned tube rows	Nu	f	Nu	f	Nu	f
2	99.3	0.321	99.3	0.321	99.3	0.321
3	99.6	0.319	99.6	0.319	99.6	0.319
4	99.8	0.313	99.8	0.313	99.8	0.313
5	100.01	0.311	100.01	0.311	100.01	0.311
6	100.3	0.31	100.3	0.31	100.3	0.31
Finned tube length: 0.5 m						
	Finned tubes per row					
	2		3		4	
Finned tube rows	Nu	f	Nu	f	Nu	f
2	99.3	0.321	99.3	0.321	99.3	0.321
3	99.6	0.319	99.6	0.319	99.6	0.319
4	99.8	0.313	99.8	0.313	99.8	0.313
5	100.01	0.311	100.01	0.311	100.01	0.311
6	100.3	0.31	100.3	0.31	100.3	0.31
Finned tube length: 0.05194 m						
	Finned tubes per row					
	2		3		4	
Finned tube rows	Nu	f	Nu	f	Nu	f
2	99.3	0.321	99.3	0.321	99.3	0.321
3	99.6	0.319	99.6	0.319	99.6	0.319
4	99.8	0.313	99.8	0.313	99.8	0.313
5	100.01	0.311	100.01	0.311	100.01	0.311
6	100.3	0.31	100.3	0.31	100.3	0.31

table, Nusselt number and friction factor are presented at different finned tube configurations from 2 finned tubes per row and 2 finned tubes rows (red polygon in Figure 4) until reach 4 finned tubes per row and 6 finned tubes rows (gray dashed box, Figure 4). Evaluations are done for tube lengths of 1 m, 0.5 m y 0.05194m as shown in table 3. Predictions show constant values of Nusselt number and friction factor for the same finned tube rows. So, the Nusselt number varies from 100.3 to 99.3 and the friction factor changes from 0.31 to 0.321 for different finned tube rows (2 to 6) of the small heat recovery. These results exhibit independence of predictions in dimensionless parameters with finned tube length and finned tubes per row in a heat recovery. The only variation of results is for configurations at different finned tube rows as shown in table 3. The main reason of dimensionless parameters variation is temperature reached by the gas phase (outside flow). These temperature variations slightly affect results because the gas phase is cooled in 3 ºC temperature difference for configurations analyzed at different finned tube rows. The results exhibit variations lower than 1% for Nusselt number and 3.6% for friction factor.

These deviations are calculated for different finned tube rows, which is not representative of the same physical phenomenon because different finned tube layouts represent different gas cooling. In spite of this situation, the results show close values in Nusselt number and friction factor. Therefore, assumptions considered in the methodology are correct because finned tube bank performance is independent of finned tube rows, number of finned tubes per row, and finned tube length in a finned tube bank. This methodology allows numerical studies in heat recoveries at industrial scale because the computational domain can be reduced in 99% as shown intable 2. So, the calculation times are finites due to the analysis is focused in a single isolated finned tube module in the fully developed flow.

4.2. Numerical Results

The predictions for velocity field, pressure field, and temperature field are shown in Figure 6. In all figure s, profiles of the variables are presented on the x-z plane, because this is the plane that exhibits most changes in properties. The velocity contours (Figure 6a) reflect an apparently symmetric behaviour but the field is slight asymmetry due to the fin helical layout. The recirculation zones are observed at the rear portion of the tubes, with reference to air flow direction. These recirculation zones are narrow because the flow tends to stick to the contour of finned tubes by turbulence generated. The recirculation region has amplitude and length of approximately 49.85 mm and 52.61 mm, respectively. The flow tends to accelerate in the free zones where the cross-section area is smaller (central region of computational domain and at the side portions of the central finned tube) with a maximum velocity of 5.017 m/s. Finally, the backwater area where the flow is stopped abruptly at the central portion of the finned tube exhibit has a width and a length of about 9.75 mm and 6.57 mm, respectively.

The pressure contours (Figure 6b) indicate an apparently symmetric pressure profile, similar to the velocity profile, which is not symmetric due to the helical fin. The results show a high-pressure zone at the front portion of the tube, taking the flow direction as the point of reference. This high-pressure zone is created due to flow stopping abruptly at the central portion of the finned tube, producing a backwater zone. At the rear of the finned tube, there is a low-pressure area that is stratified at the outlet of single isolated module. The mean pressure at the inlet and outlet of the module is 23.184 Pa and 9.019 Pa, respectively, as shown in table 4. The pressure drop is 14.165 Pa, which represents a friction factor of 0.311 (table 4). The deviation of numerical results is 3.13% and 0.32% for pressure drop and friction factor, respectively. So, numerical predictions with boundary conditions calculated from methodology correspond to results obtained with Weierman´s [15] model.

Numerical predictions of temperature contours (Figure 6c) show an apparently symmetric profile, which are not symmetric due to the helical fin. The results exhibit that the high-temperature region is located at the central region of the computational domain, whereas the low-temperature areas are located at the rear of the finned tubes, viewed with respect to the flow direction. So, the most important heat transfer effects occur at the front of the extended surfaces and at the sides of the finned tube between the backwater zone and the recirculation zone. This pattern is attributed to the air flow coming into abrupt contact with the finned tubes, and to turbulence being created at the sides of the finned tubes. On the other hand, the temperature contours exhibit that flow temperature is dominated for inside fluid temperature in the recirculation zone. The mean temperature at inlet and outlet of module is 43.76°C and 34.73°C (table 4). The temperature difference is 9.03°C, which represents a numerical-mean Nusselt number of 95.1, as shown in table 4. Numerical results show deviations of 4.25% and 5.18% for temperature difference and Nusselt number, respectively. These results show that numerical predictions are close to results obtained with Kawaguchi´s et al. [18] model. So, the numerical simulation represents adequately heat transfer in a single isolated finned tube module with boundary conditions calculated from methodology.

Table 4. Comparative results.

	Numerical Simulation	model
		Weierman
P_{inlet} (Pa)	24.44	na
P_{outlet} (Pa)	10.27	na
ΔP (Pa)	14.17	13.74
deviation	3.13%	reference
f	0.311	0.31
deviation	0.32%	reference
		Kawaguchi
T_{inlet} (ºC)	40.78	na
T_{outlet} (ºC)	31.78	na
ΔT (ºC)	9.0	9.4
deviation	4.25%	reference
Nu	95.1	100.3
deviation	5.18%	reference

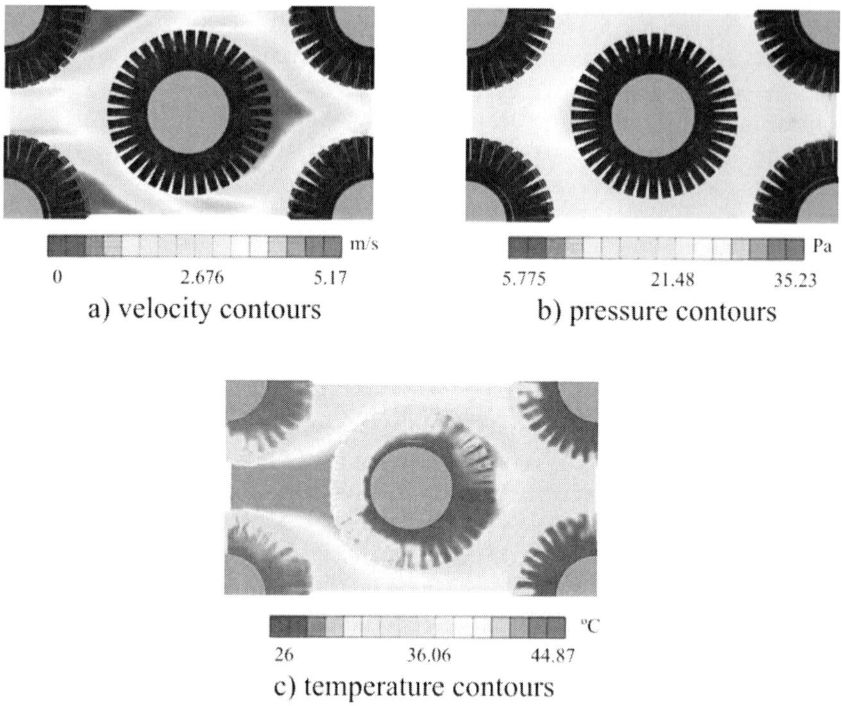

Figure 6. Numerical predictions.

5. CONCLUSION

The methodology, based on the LMTD method, for calculating boundary conditions in a single isolated finned tube module is adequate because the Nusselt number and friction factor show a quasi-constant behaviour for different finned tube length and finned tubes per row. The maximum deviations for Nusselt number and friction factor are lower than 1% and 3.6%, respectively. Therefore, the methodology can be used in numerical analysis of heat recoveries at industrial scale because the computational domain can be reduced in 99% which allows finite computational times. Numerical predictions in the single isolated finned tube module show that pressure contours are adequate because mean pressure drop and mean friction factor exhibit a deviation of 3.13% and 0.32% with respect to model developed by Weierman [15]. On the other hand, the pressure contours exhibit that the high pressure values are located in backwater zone and the main dissipative effect of flow energy is located at this region. Therefore, the pressure drop is dominated for backwater zone. The temperature field show a deviation of 4.25% and 5.18% for temperature difference and Nusselt number with respect to Kawaguchi´s, et al [18] model. The temperature contours exhibit that bare tube temperature is dominated by inside fluid temperature. Therefore, the inside fluid temperature must be considered as a lower limit temperature for cooling of flue gases in order to avoid dew point of acid gases.

ACKNOWLEDGEMENT

We appreciate the support given to the research presented here by Universidad Nacional Autonoma de Mexico (Direccion General de Asuntos del Personal Academico, PAPIIT- IN106112-3), and Universidad Autonoma Metropolitana Azcapotzalco.

REFERENCES

1. S. B. Beale, D. B. Spalding, 1999A Numerical Study of Unsteady Fluid Flow in In-line and Staggered Tube Banks. Journal of Fluids and Structures, Vol.13, No.6, pp. 723-754, ISSN 0889-9746
2. G. Comini, G. Croce, 2003Numerical Simulation of Convective Heat and Mass Transfer in Banks of Tubes. International Journal for Numerical Methods in Engineering, Vol.57, No.12, pp. 1755-1773, ISSN 1755-1773
3. S. B. Beale, 2007Use of Streamwise Periodic Boundary Condition for Problems in Heat and Mass Transfer. ASME Journal of Heat Transfer, Vol.129, No.4, pp. 601-605, ISSN0022-1481
4. S. B. Beale, 2008Benchmark Studies for the Generalized Streamwise Periodic Heat Transfer Problem. ASME Journal of Heat Transfer, Vol.130, No.11, pp. 114502_1-114502_4, ISSN0022-1481
5. S. Benhamadouche, D. Laurence, 2003LES, Coarse LES, and Transient RANS Comparisons on the Flow Across a Tube Bundle. International Journal of Heat and Fluid Flow, Vol.24, No.4, pp. 470-479, ISSN 0142-727X.
6. M. Salinas-Vázquez, M. A. de la Lama, W. Vicente, E. Martínez, 2011Large Eddy Simulation of a Flow through Circular Tube

Bundle. Applied Mathematical Modelling, Vol.35, No.9, pp. 4393–4406. ISSN 0307-904X.

7. R. Hofmann, K. Ponweiser, 2008Experimental and Numerical Investigations of Serrated-Finned Tubes in Cross-Flow. Available from:www.zid.tuwien.ac.at/fileadmin/files_zid/projekte/2008/08-302-2.pdf.

8. S. R. Mcilwain, 2010A Comparison of Heat Transfer Around a Single Serrated Finned Tube and a Plain Finned Tube. IJRAAS, Vol.2, No.2, pp. 88-94, ISSN

9. S. R. Mcilwain, 2010A CFD Comparison of Heat Transfer and Pressure Drop Across Inline Arragement Serrated Finned Tube Heat Exchangers with an Increasing Number of Rows. Vol.4, No.2, pp. 162-169, ISSN

10. A. Lemouedda, A. Schmid, E. Franz, M. Breuer, A. Delgado, 2011Numerical Investigations for the Optimization of Serrated Finned-Tube Heat Exchangers. Applied Thermal Engineering, Vol.31, No. 8-9, pp. 1393-1401, ISSN 1359-4311

11. V. Ganapathy, 2002Industrial Boilers, Heat Recovery and Steam Generators: Design, Applications and Calculations. In Marcel Dekker, (Ed), 333507 ISBN 0-82470-814-8York.

12. V. Gnielinski, 1976New equations for heat and mass transfer in turbulent pipe and channel flow. Int. Chem. Eng. Vol.16, No.1, pp. 359-366, ISSN 0020-6318

13. M. V. Rane, S. Tandale, 2005Water-to-water heat transfer in tube-tube heat exchanger: Experimental and analytical study. Applied Thermal Engineering, Vol. 25 No.17 , pp. 2715-2729, ISSN 1359-4311

14. A. Bejan, 1995Convection Heat Transfer. In Wiley, (Ed), pp. 391-395, ISBN 0-471-27150-0

15. C. Weierman, 1976Correlations Ease The Selection of Finned Tubes. Oil and Gas Journal, Vol.74, No.36, pp. 94-100, ISSN 0030-1388.

16. ESCOA Turb-X HF Rating Instructions.1979Extended Surface Corporation of America (ESCOA), Pryor, OK.

17. A. Nir, 1991Heat Transfer and Friction Factor Correlations for Crossflow over Staggered Finned Tube Banks. Heat Transfer Engineering, Vol.12, No.1, pp. 43-58, ISSN 0145-7632

18. Kiyoshi Kawaguchi, Kenichi Okui, Takaharu Kashi.2005Heat transfer and pressure drop characteristics of finned tube banks in forced convection. Journal of Enhanced Heat Transfer, Vol.12, No.1, pp. 1-20, ISSN 1065-5131

19. E. Martinez, G. Soto, W. Vicente, M. Salinas, 2005Comparative Analysis of Heat Transfer and Pressure Drop in Helically Segmented Finned Tube Heat Exchangers. Applied Thermal Engineering, Vol.30, No.11-12, pp. 1470-1476, ISSN 1359-4311.

20. Weale. P. E. Jonh, H. Peter, P. E. Rumsey, Sartor. P. E. Dale, Eng. Lee, Lock, 2002Laboratory Low-Pressure Drop Design. ASHRAE Journal, Vol. August Vol. August pp. 38-42. ISSN 0001-2491

21. E. Martinez, W. Vicente, M. Salinas, G. Soto, 2010Thermal Design Methodology of Industrial Compact Heat Recovery with Helically Segmented Finned Tubes. Heat Exchangers; design, types and applications. In Nova Publishers, (Ed), Series: Energy Science, Engineering and Technology, pp. 215-228. ISBN: 978-1-61761-308-

22. E. Martínez, W. Vicente, M. Salinas, G. Soto, 2009Single-phase experimental analysis of heat transfer in helically finned heat exchangers, Applied Thermal Engineering, Vol. 29No. 11-12, pp. 2205-2210, ISSN 1359-4311

23. A. Favre, (969, Problems of Hydrodynamics and Continuum Mechanics, SIAM.

24. V. Yakhot, S. Orszag, 1986Renormalization Group Analysis of Turbulence. Basic Theory. Journal of Scientific Computing, Vol.1, No.1, pp. 3-51, ISSN 08857474.

25. S. V. Patankar, C. H. Liu, E. M. Sparrow, 1977Fully Developed Flow and Heat Transfer in Ducts Having Streamwise-Periodic Variations of Cross-Sectional Area, ASME Journal of Heat Transfer, 99, pp180-186

26. K. M. Kelkar, S. V. Patankar, 1987Numerical Prediction of Flow and Heat Transfer in a Parallel Plate Channel with Staggered Fins, ASME Journal of Heat Transfer, 109, pp 25-30

27. E. E. Martínez, 2011Simulación Numérica de un Flujo de Gases Turbulentos en un Banco de Tubos aletdos en Geometría Compleja. Ph.D Thesis. Universidad Autónoma Metropolitana, México City, México.

28. J. C. Ludwing, H. Q. Qin, D. B. Spalding, 1989The PHOENICS Reference Manual. Technical Report CHAM TR/200, CHAM Ltd, London.

29. Y. Choi, J. Hong, H. Hwang, J. Choi, 2008Cartesian Grid Method with Cut Cell in the Mold Filling Simulation. Journal Materials Science British Technology, Vol.24, No.3, pp. 379-382, ISSN 1005-0302.

2

CHAPTER

A 2D Inverse Problem of Predicting Boiling Heat Transfer in a Long Fin

Tadeusz Orzechowski1

Environmental Engineering Department, EcoClimatica Association, Kielce University of Technology, Aleja Tysia̜clecia Pan´stwa Polskiego 7, 25 -314 Kielce, Poland

ABSTRACT

A method for the determination of local values of the heat transfer coefficient on non-isothermal surfaces was analyzed on the example of a long smooth-surfaced fin made of aluminium. On the basis of the experimental data, two cases were taken into consideration: one-dimensional model for $Bi < 0.1$ and two-dimensional model for thicker elements. In the case when the drop in temperature over the thickness could be omitted, the rejected local values of heat fluxes were calculated from the integral of the equation describing temperature distribution on the fin. The corresponding boiling curve was plotted on the basis of temperature gradient distribution as a function of superheat. For thicker specimens, where $\mathrm{Bi} > 0.1$, the problem was modelled using a 2-D heat conduction equation, for which the boundary conditions were posed on the surface observed with a thermovision camera. The ill-conditioned inverse problem was solved using a method of heat polynomials, which required validation.

1 INTRODUCTION

Heat transfer in the nucleate boiling regime, especially the heat flux density, has been a subject of interest since the 1950s. Although much research has been devoted to the problem, no satisfactory explanation of the physical phenomena accompanying the process has been found. That is mainly due to the fact that the physics of mass, momentum and energy transfer between the surface of the solid body and the wetting fluid is difficult to explain and describe. Another reason is the nucleation, growth and detachment of vapour bubbles, which, at appropriately high wall superheat, lead to a crisis called the first critical point.

A wide review of physical models describing the boiling process and a means of predicting the operation of smooth surface heat exchangers were presented by Dhir [1]. In the last 80 years, a number of empirical and mechanism based correlations have been proposed for nucleate pool boiling. Empirical correlations differ from each other substantially with respect to the functional form of the heat flux density depending on fluid and surface properties. Additionally, all of them require knowledge of the density of active sites, bubble diameter, its frequency of departure, etc. However, because of the complex nature of the coexisting processes, it has not been possible to develop comprehensive formulas for these parameters.

The experimental data concerning boiling heat transfer described in the literature focus on isothermal surfaces. This is primarily due to the measuring technique, which involves mounting a specimen on the heating element. The power is determined directly by measuring the electricity supplied, as done by Ahn et al. [2,3] and Naphon et al. [4], or indirectly by applying a gradient method including measurement of temperature at several points along the specimen axis, as done by Ahn et al. [5]. The final result of the experiment is a boiling curve, which shows the relationship between the heat flux density and the heat transfer coefficient as a function of wall superheat. The generalized results are illustrated in the form of correlations for the Nusselt number.

In the case of nucleate boiling heat transfer on smooth surfaces, the Rohsenow equation is used, and values of the required constants are selected for various surface-fluid combinations, Pioro [6] and Pioro et al. [7]. The factors involved in boiling heat transfer include the type, state, shape and orientation of the surface, the type of the fluid, possibly the amount of foreign particles, etc. However, the correlation relationships determined by different researchers even under the same or similar geometric and material conditions will differ. For instance, Kolev [8] analyzes a number of different correlation equations describing heat transfer for water boiling at ambient pressure on smooth surfaces. The values of the heat flux density calculated by means of these relationships differ, which can be explained by the fact that the measurements were conducted under different conditions and the surfaces differed in state. Sixteen different correlations to predict the boiling heat transfer coefficient are analyzed and compared to evaluate their use in evaporator design by Adib and Vasseur [9]. From this review, it appears that it is difficult to predict the proper value of the heat flux on the basis of the existing formulas found in the literature. A large number of correlations available in the literature are seldom satisfactory. Using the available literature and experimental test results, some researchers try to find a universal correlation for the boiling heat transfer coefficients, as Shah [10]. Although they make progress in prediction accuracy, so far such attempts are still not satisfactory for general purposes.

Physical phenomena accompanying nucleate boiling are also considered theoretically. A 2D numerical investigation of boiling with a energy balance simulation is applied by Orzechowski et al. [11]. Measured and numerically calculated values of heat transfer coefficient demonstrate good congruence, which indicates a correct selection of measurement methods and a cogent analysis of the experimental results. The application of the heat transfer theory to pool boiling is discussed by Roh [12]. The study provides an explicit illustration of the kinetics of phase transition mechanisms at the interface of two different phase materials.

Considerable divergence of results discussed in the literature indicates that the designed heat transfer elements should be measured separately. This is particularly essential in the case of non-isothermal surfaces, where the boiling heat transfer processes differ significantly from those observed on surfaces with uniformly distributed temperature, as shown by Orzechowski [13]. A typical example, discussed by Krikkis [14], Liaw et al. [15] and Gonzalez-Fernandez et al. [16], is a fin, on the surface of which different boiling regimes can be distinguished. Similar conclusions are sought analytically by Turkyilmazoglu [17] and Kundu et al. [18]. They determine the performance of different fin geometries by analyzing the temperature-dependent thermal conductivity of the fin material, the variable heat transfer coefficient under different boiling regimes or surface conditions. A original technique for determining the transient heat transfer coefficients from measured temperatures inside solids is presented in [19]. Numerical examples are given as verification of the method.

Maintaining the heat transfer rate and the desired operating temperature is an important task when a reliable operation of machinery components is to be provided. The proper design of the heat exchangers can produce an efficient solution to the problem. Chen et al. [20] applied an inverse method together with FLUENT software and the experimental temperature distribution to investigate the non-uniform heat transfer coefficient on the fin in the heat sink for various fin spacings. Torabi and Aziz [21] investigated the thermal performance of T-shaped fins. They accounted for temperature-dependent thermal conductivity, surface emissivity, convection heat transfer coefficient, and non-zero convection and radiation sink temperatures. The problem was solved numerically. Konda Reddy et al. [22] presented a methodology for the estimation of temperature-dependent heat transfer coefficient for a vertical rectangular fin operating in a steady state under natural convection regime. Heat transfer coefficient was considered as a power law function of temperature excess. Thermal performance of variable diameter micro-pin fins was calculated by Díez et al. [23]. On the basis of an exact solution of smooth pin fins for three selected profiles (hyperbolic, trapezoidal and concave parabolic ones), the approximate series solution was given. The effect of surface roughness on efficiency increase was evaluated. Three different types of fins made of aluminium and copper, at a fixed fin array volume were discussed by Huang et al. [24] while looking for the most efficient solutions to heat sinks. Results obtained by using the Levenberge-Marquardt method to solve 3-D inverse problem were based on the numerical experiments. Estimation of unknown parameters in a rectangular fin satisfying a predefined temperature are discussed by Das et al. [25]. The study is useful for selecting various unknowns in a conductive-convective and radiative fin for required temperature and efficiency.

Grysa et al. [26] presented an approximate method of solving nonlinear inverse stationary heat conduction identification of boundary condition. The obtained results were compared with the direct solution of the problem. A numerical study of the inverse problem of determining time-variable surface heat flux in a plane wall, with constant or temperature dependent thermal properties, is given by Zueco et al. [27]. Direct and inverse heat conduction problems are solved with the network simulation method. Both the accuracy and effectiveness of the method has been discussed.

Considerations such as those mentioned above, aim to determine the performance and optimum design of fin assemblies. The aim of this study was to derive an efficient algorithm for the determination of local values of the boiling heat transfer coefficient using a specimen made of aluminium.

2 EXPERIMENTAL PROCEDURE AND MEASUREMENTS

Investigations were conducted at the test facility, specially arranged for the task. The diagram of the facility is presented in Fig. 1.

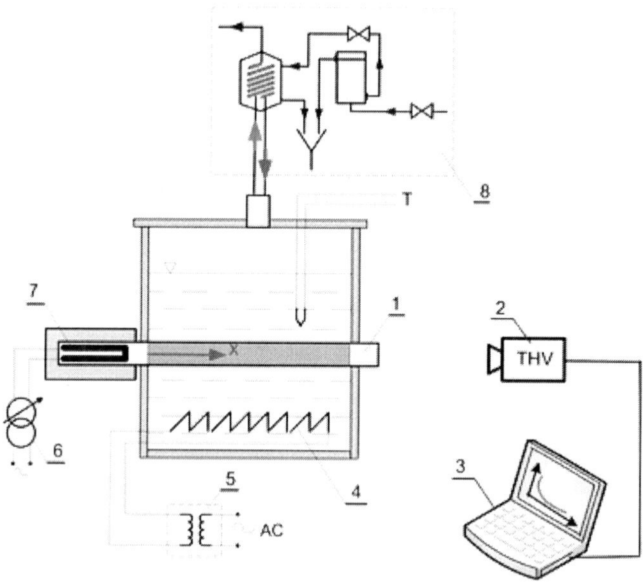

Figure 1. Diagram of the test facility for non-isothermal specimens: *1* examined element, *2* thermovision camera, *3*data acquisition system, *4* auxiliary heater, *5* separation system with the control of supplied electric power, *6*autotransformer with the electric power measurement, *7* main heater, *8* cooling and condensate recovery system, *T* thermocouple

The experiment involved a long fin-shaped specimen with a width of 12 mm, an active length of 90 mm and a different thickness, which was placed between two textolite plates. One of the specimen surfaces was in contact with the boiling fluid, and the other was surrounded by quiescent atmospheric air. The test facility was equipped with two independent heating systems. The main heating element, whose power was controlled with an autotransformer, was located at the fin base. The auxiliary heating system, made of spirally coiled resistance wire, was positioned on the bottom of the vessel. A constant supply of the electric power to the auxiliary system was maintained throughout all measurement series. The task of the auxiliary system was to sustain the boiling process in the whole volume of the liquid. Supplying power to the auxiliary system was preceded by a measurement intended to examine the effect produced by the power supply on the results of investigations. Within the whole range of the power supply, no such effect was observed. The test facility was equipped with a cooling and condensate recovery system. Vapours of the boiling liquid condensed in the system of two tap water-cooled condensers, and then flowed, by gravity, into the vessel.

The outer surface of the examined element was observed with a thermal imaging camera, as shown in Fig. 2. Since high emissivity of the observed surface is required to reduce the uncertainty of the IR temperature measurement, the outer surface was coated in black paint. The optical homogeneity of the observed surface and the emission coefficient in the camera spectral range were determined during the initial calibration process. The experiments were performed using a VIGO System V50 camera, operating in the long-wave infrared range (8–11 µm), equipped with a 384 × 288 microbolometer FPA (Focal Plane Array), whose sensitivity at a temperature of 30 °C is <0.08 K. The main heater was mounted at the base of the specimen, where the predetermined time-invariant electric power was supplied. The heater outer surface was shielded so that the emitted radiation would not affect the temperature measurement. The tests were conducted for refrigerant FC 72 at ambient pressure, which after evaporation was directed onto the cooling system in order to recover the condensate. The specimen had to be adequately fixed so that the temperature field observed with the thermal imaging camera could be treated as a one-dimensional quantity. The result of the measurement was the distribution of temperature along the length for the central line of the fin. The test was carried out during refrigerant FC 72 boiling at ambient pressure on a non-isothermal smooth surface of an aluminium fin. Exemplary measurement results for a smooth surface specimen 3 mm in thickness are shown in Fig. 2. The results were obtained for three different heating powers supplied to the fin base.

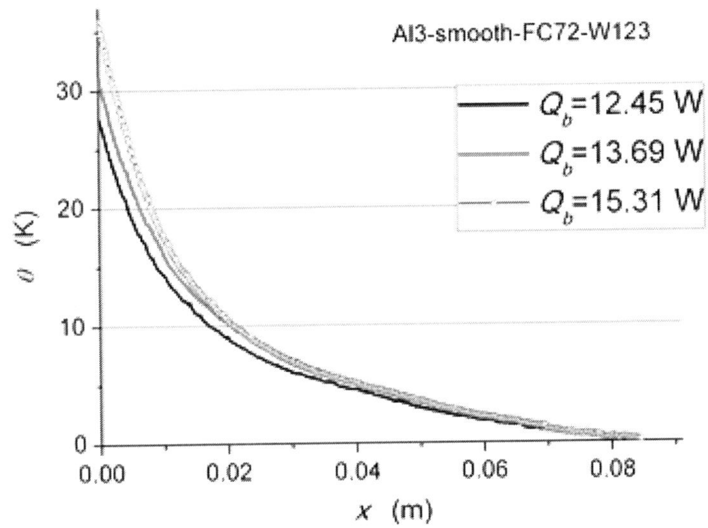

Figure 2. Axial temperature distribution for a 3 mm thick smooth aluminium fin

Measurements of the electrical power applied to the main heater do not reflect the heat dissipated by the fin. It is thus difficult to estimate the amount of heat lost to the surroundings. Moreover, the changes are uneven as they are dependent on the changes in load. It can be assumed that for sufficiently thin specimens the Biot number is small, i.e. Bi < 0.1, since there is no change in temperature with thickness, it can be omitted and the total heat conducted at the base of the fin $Q_b$can be calculated using Fourier's law. The necessary temperature gradient can be determined from the measurement data shown in Fig. 2.

The same experimental set-up was used for a 12 mm thick smooth specimen made of aluminium. It had identical chemical composition and surface morphology as the 3 mm specimen. The electrical power supplied to the main heater was adjusted to keep the temperature of the outer surface at the fin base as close as possible to that reported for the thinner sample. Figure 3 shows the axial temperature distributions in the two fins.

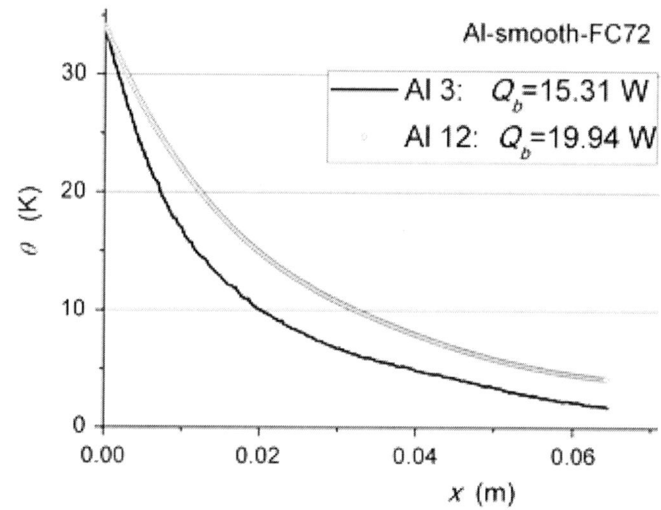

Figure 3. Temperature distribution along the centerline of aluminium fins of different thickness

Direct observation of the boiling process at the fin surface shows that for the two samples heat transfer at the base occurs in the fully developed nucleate boiling regime in the vicinity of the critical point. The heat transfer coefficient for nucleate boiling is frequently described by the Rohsenow correlation, which indicates its exponential dependence on the wall superheat [6]. In this case, the total amount of heat dissipated by the fin with a constant cross-section is proportional to the square root of the surface at the base, i.e. Qb~Fb√Qb~Fb (see Ref. [28]). The heat transfer surfaces of both fins are of equal dimensions, have the same spatial orientation and morphology, and fins differ only in thickness. With the assumption of constant temperature in the fin cross-section, the amount of heat dissipated by the 12 mm thick fin should be twice as high as the amount of heat dissipated by the 3 mm thick fin. Thus, according to Fourier's law, the derivative of the temperature at the base should be twice as small as that for the thicker fin. However, the values measured were different; they were −2412.8 and −1519.1 K/m for the thinner and thicker fins, respectively, while their ratio was 0.63. This large discrepancy indicates that the 1-D analysis cannot be used for fins with a relatively large thickness. Thus, the problem must be solved numerically.

3 PROBLEM FORMULATION AND THE METHOD OF SOLUTION

The distribution of temperature on the outer surface measured with the thermovision camera is the evidence of the heat transfer taking place on the surface invisible to the camera. As shown in Fig. 3 the surface temperature distributions along the centrelines of fins of 3 and 12 mm, however, are not the same, which results only from different thicknesses of the samples. The different slopes of the temperature distribution for both specimens indicate that a thicker specimen does not satisfy the Biot number condition and 1-D approximation cannot be applied automatically.

3.1 One-dimensional inverse problem

The method for determining a local value of the heat transfer coefficient over a non-isothermal fin surface can be used only with one-dimensional systems, i.e. ones where the drop in temperature over the thickness can be omitted. This assumption is true for numbers $Bi < 0.1$. At nucleate boiling regime, for extended surfaces, the heat transfer coefficients reach values even higher than 0.1 MW/m^2K. Thus, using a one-dimensional model is possible only for elements with a small thickness that are made of high thermal conductivity materials. For such large values of heat transfer coefficients, a copper fin <0.4 mm in thickness is required to strictly satisfy the Biot number criterion. Boiling heat transfer of FC72 on the smooth surface of the aluminium sample requires that the sample thickness should be a little below 4 mm.

The performance of a straight fin with a constant cross-section is analysed on the basis of one-dimensional expression for the temperature distribution:

$$\frac{d^2\theta}{dx^2} = m_1^2 q_\alpha = m_1^2 \alpha\theta , \tag{1}$$

where $m_1^2 = P/\lambda F$, with P denoting the perimeter of the fin cross-section and F the cross-sectional area, λ thermal conductivity, q_α local value of the heat flux rejected by the fin lateral side, α heat transfer coefficient.

In inverse problems, the temperature distribution is also used to determine the heat transfer coefficient and the heat flux density. The right-hand side of Eq. 1 is dependent only on the wall superheat. Thus, the integration gives

$$\left(\frac{d\theta}{dx}\right)^2 = 2m_1^2 \int q_\alpha d\theta + C = b_0 + b_1\theta + b_2\theta^2 + b_3\theta^3 + b_4\theta^4 + \cdots \tag{2}$$

where C is an integration constant calculated, if necessary, from the boundary condition.

Assuming that the result of integration of the right-hand side of Eq. (2) is described by the holomorphic function, its value can be represented by a power series as shown in (2).

Differentiation of both sides of Eq. (2) is replaced by the following expression for the heat flux density:

$$q_\alpha = \frac{1}{2m_1^2}\left(b_1 + 2b_2\theta + 3b_3\theta^2 + 4b_4\theta^3 + \cdots\right) \tag{3}$$

where m21m12 is defined in Eq. (1) while coefficients b_i result from polynomial fit (see Fig. 5).

3.2 Formulation and solution of a 2-D ill-posed problem

In the case of thicker fins, where the differences in temperature between the outer and inner surfaces are large, the assumption of a one-dimensional model leads to considerable errors. A good solution is to apply a method of non-invasive measurement to determine the temperature field on the accessible surface and calculate the heat transfer surface area. Then, the local values of the heat transfer coefficient and heat flux are established.

The conditions of the measurement described above require that the heat transfer coefficients should be determined on the basis of at least the two-dimensional Laplace equation for OABC area shown in Fig. 4.

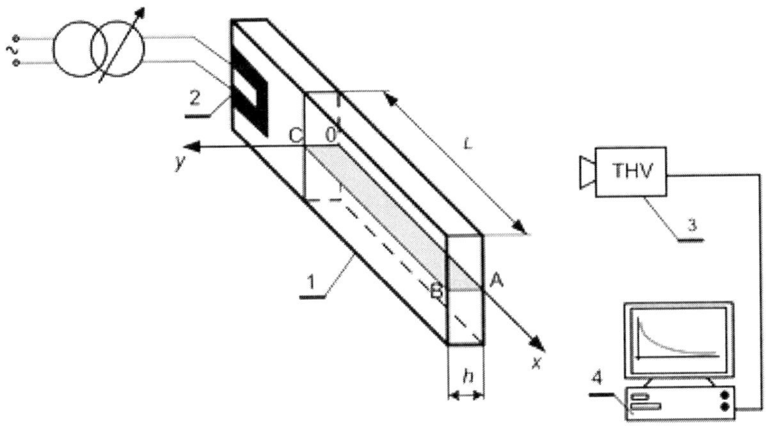

Figure 4. View of the element measured: *1* fin, *2* main heater, *3* thermovision camera, *4* data acquisition and processing station, *L*, *h* length and thickness of the measured section, respectively

Two of the boundary conditions on the side observed by the thermovision camera, i.e. the temperature distribution and the dissipated heat flux, are generally easy to estimate. The other boundary conditions need to be determined individually. Thus, that constitutes not only an inverse but also an ill-posed problem. The problem described in this study was solved using heat harmonic polynomials (sometimes called heat polynomials). This method, proposed by Ciałkowski et al. [29], makes it necessary to find an exact solution in the form of an approximation satisfying the two-dimensional heat conduction equation and approximately satisfying the boundary conditions.

For a fin operating in the system presented in Fig. 4, the modelling was based on the stationary two-dimensional heat conduction equation,

$$\frac{\partial^2 \theta}{\partial x^2} + \frac{\partial^2 \theta}{\partial y^2} = 0 \tag{4}$$

θ the wall superheat.

It is expected that the solution to θ(x,y) should be looked for in the OABC area. Heat is removed from the side surface, BC, at pool boiling of the surrounding medium. The aim of the calculations is to determine some local values of the heat flux density of this energy. The four boundary conditions required to solve the problem are:

- for the thermovision camera OA side (see Fig. 4):

$$\theta(x_m, 0) = f(x_m) \tag{5}$$

$$\frac{\partial \theta}{\partial y} = 0 \tag{6}$$

where x_m is the coordinate of the experimental points. Condition (5) represents the distribution of discrete surface temperature measured along the central line of the fin, OA, at M points with the x_m coordinates. Additionally, it is assumed that the heat flux rejected to the surroundings along OA is negligible in comparison with the flux generated from the BC surface. The phenomenon of boiling heat transfer is observed along the BC side of the fin, while the OA edge is in contact with quiescent atmospheric air. The heat transfer coefficient is several orders of magnitude lower for OA in comparison with BC. The estimations show that for refrigerant boiling at atmospheric pressure the heat flux dissipated by OA is <1 % of the heat flux dissipated by BC, which confirms the correctness of assumption (6). Moreover, it is a condition of the fin central symmetry.

- The temperature distribution in the fin cross-section at the base, OC, is assumed to be constant:

$$\theta(0, y) = \theta_b = const. \tag{7}$$

- For the fin top (the AB side in Fig. 4), it is assumed that:

$$\frac{\partial \theta(L, y)}{\partial x} = 0 \tag{8}$$

where θ_b is the wall superheat at the fin base. Satisfying condition (8) requires using a fin with a sufficient length, *L*. It should be noted that the condition on the BC side is not known, whereas for $y = 0$, two conditions are given.

The problem analyzed here is not only inverse but also ill-posed. To solve it, a method of resolving functions (in this case, some kind of polynomials for the local temperature) needs to be employed. For a given differential equation, it is always necessary to determine a generating function dependent on the parameter p. In accordance with [29], the following function was used:

$$f = e^{p(x+iy)}, \quad \text{where } p \in R \tag{9}$$

because it satisfies the differential Eq. (4).

The coefficients of the Taylor series expansion of the above function relative to the parameter p are the desired polynomials solving the differential equation. The series expansion of the exponential function (9) is given by the following formula:

$$e^{p(x+iy)} = \sum_{j=0}^{\infty} (x+iy)^j \frac{p^j}{j!} \tag{10}$$

This gives two sequences of harmonic polynomials, the so-called heat polynomials, one being a real part and the other an imaginary part of the complex function, $(x + iy)^j$:

$$\begin{aligned} u_j(x,y) &= \mathrm{Re}\left[(x+iy)^j\right] \\ w_j(x,y) &= \mathrm{Im}\left[(x+iy)^j\right] \end{aligned} \tag{11}$$

The same order polynomials u and w are used interchangeably to obtain a sequence of harmonic polynomials V_j, each of which satisfies the Laplace equation and can be treated as a particular solution of Eq. (4):

$$\begin{aligned} V_0 &= u_0 = 1 \\ V_1 &= u_1 = x \\ V_2 &= w_1 = y \\ V_3 &= u_2 = x^2 - y^2 \\ V_4 &= w_2 = 2xy \\ V_5 &= u_3 = x^3 - 3xy^2 \\ V_6 &= w_3 = 3x^2y - y^3 \\ &\ldots\ldots\ldots\ldots\ldots\ldots \end{aligned} \tag{12}$$

Its general solution is approximated by means of a linear combination of the heat polynomials, which constitute the approximation *Θ(x,y)* of the exact solution having the form of:

$$\begin{aligned}\theta(x,y) \approx \Theta(x,y) &= \sum_{j=0}^{N} c_j \cdot V_j(x,y) \\ &= c_0 + \sum_{j=0}^{N/2} c_{2j+1}\left(x \cdot u_{2j} - y \cdot w_{2j}\right) \\ &+ \sum_{j=0}^{N/2-1} c_{2j+2}\left(x \cdot w_{j+1} + y \cdot u_{j+1}\right)\end{aligned} \tag{13}$$

c_j coefficients, *N* degree of the polynomial approximating the exact solution.

Expression (13) satisfies the Laplace Eq. (4); in a general case, however, it does not exactly satisfy the boundary conditions. One can only demand that the acceptable error be as negligible as possible. The condition assumed in this way allows us to select unknown coefficients c_j. It is thus essential to express the error functional as:

$$\begin{aligned}I = &\frac{1}{M \cdot \theta_b^2} \sum_{m=1}^{M} \left[\Theta(x_m,0) - f(x_m)\right]^2 + \frac{L}{\theta_b^2} \int_0^L \left[\frac{\partial \Theta(x,0)}{\partial y}\right]^2 dx \\ &+ \frac{1}{L \cdot \theta_b^2} \int_0^h \left[\Theta(0,y) - \theta_b\right]^2 dy + \frac{L}{\theta_b^2} \int_0^h \left[\frac{\partial \Theta(L,y)}{\partial x}\right]^2 dy \rightarrow \min\end{aligned} \tag{14}$$

where *M* is the total number of measured points.The following terms of the above error functional refer to the differences in boundary conditions (5)–(8) and value of function (13) which represents the approximation of the exact solution of Laplace Eq. (4).All components required in the above expression should be determined from the assumed form of the approximation (13). Then, by differentiating functional (14) in relation to all coefficients c_j, the following system of linear equations is obtained:

$$\frac{\partial I}{\partial c_i} = 0 \quad \text{for } i = 0, 2, 3 \ldots, N \tag{15}$$

Thus, the problem is solved by applying the *N* + 1 linear equation systems with unknown values of vector $C = C[c_1, c_2, \ldots, c_N]$ with unknown coefficients c_j:

$$A \cdot C = B \tag{16}$$

where

$$A = \begin{bmatrix} a_{1,1} & \cdots & a_{1,N} \\ & \ddots & \\ & a_{i,j} & \\ & & \ddots \\ a_{N,1} & & a_{N,N} \end{bmatrix}, \qquad B = \begin{bmatrix} b_1 \\ \vdots \\ b_j \\ \vdots \\ b_N \end{bmatrix} \tag{17}$$

The elements of matrices ***A*** and ***B*** are defined by the following formulae:

$$a_{ij} = \frac{1}{M} \sum_{m=1}^{M} V_j(x_m, 0)\, V_i(x_m, 0) + L \int_0^L \frac{\partial V_j(x,0)}{\partial y} \frac{\partial V_i(x,0)}{\partial y} dx$$

$$+ \frac{1}{L} \int_0^h V_j(0,y)\, V_i(0,y) dy + L \int_0^h \frac{\partial V_j(L,y)}{\partial x} \frac{\partial V_i(L,y)}{\partial x} dy = 0$$

$$b_i = \frac{1}{M} \sum_{m=1}^{M} f(x_m)\, V_i(x_m, 0) + \frac{1}{L} \int_0^h \theta_b\, V_i(0,y) dy \tag{18}$$

As a result, the desired coefficients of the linear combination of functions (12) are obtained, which, introduced in Eq. (13), form an approximation of the exact solution. By using the obtained relationships, the thermal field in the analyzed area can be determined.

4 METHOD VALIDATION

The temperature measured on the outer surface OA (see Fig. 4) provides information on the processes occurring on the inner surface of the fin, not accessible to the thermovision camera. As a result, the temperature measured is different from the actual one occurring during intensive heat transfer.

The thermal field on the fin surface is measured at many points with a thermovision camera. That makes it possible to apply typical procedures of smoothing and then numerical differentiation available in a number of calculation packages. For the 3 mm thick aluminium fin, the results of the analysis of the measurement data from Fig. 2 are shown in Fig. 5.

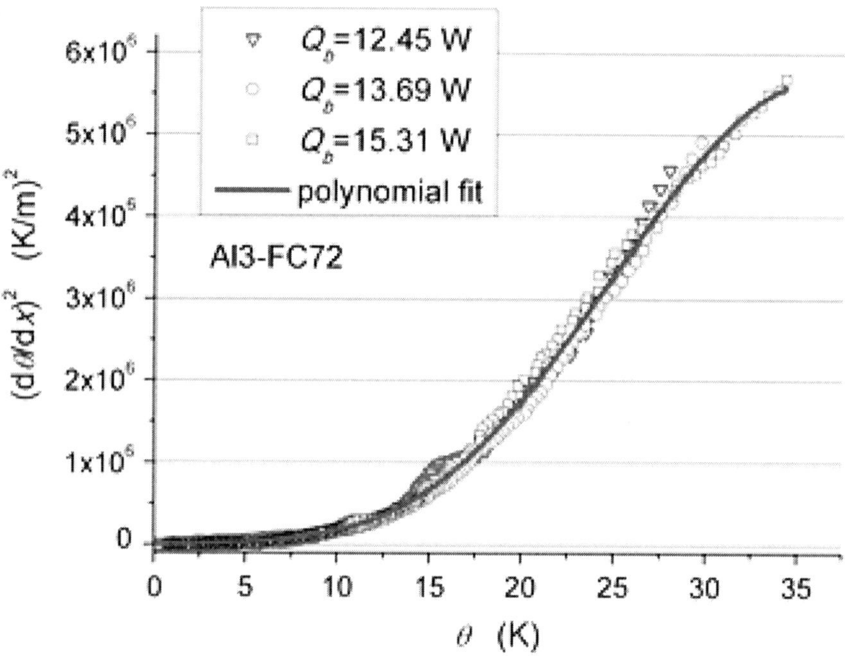

Figure 5. Square of the temperature gradient versus wall superheat

Using the polynomial approximation to the data given in Fig. 5, all coefficients of Eq. (2) could be determined and then the local heat fluxes q_α and heat transfer coefficients could also be established. The corresponding boiling curve for the investigated surface is drawn in Fig. 7. Here and hereafter, the boiling curve is defined locally, i.e. for the local values of heat flux and heat transfer coefficients from superheat. Detailed analysis of the method and its uncertainties are discussed by Orzechowski [28].

The major factor affecting the proper determination of the heat flux density is the limited accuracy of the measuring device. The smaller are the differences between the temperature values, the greater effect of the measurement accuracy is noted. In this study, a regime around the critical heat flux and free convection boiling was analyzed. The determination of the heat flux density or heat transfer coefficient from error-prone measurement data is very difficult and requires validation of the method, especially if the problems are ill-posed.

The correctness of the developed algorithm is generally verified by a numerical experiment, in which a similar problem is solved analytically. A more reliable method should use, if possible, the experimental data. For that reason, an additional independent test was conducted at the same test facility using the same devices and under the same conditions. The experiments involved measuring the distribution of temperature on the aluminium specimen with a thickness of 3 mm. For aluminium, whose conductivity is very high, and, for the small thickness of the specimen, the condition $Bi < 0.1$ was well satisfied nearly over the entire wall superheat range, and the heat transfer could be treated as a one-dimensional problem. The boiling curve was plotted using the method described in Sect. 3, and the results were compared with analogous data using the proposed method of heat polynomials.

Figure 6 shows the temperature measured on the outer surface of the aluminium fin and the temperature calculated for the inverse ill-posed problem using the method of heat polynomials with the boundary conditions (5)–(8). In Fig. 6b, a difference in the measured and calculated temperature is shown. Its mean value is close to zero, and nearly all points are located in the range ±0.2 K. Only a few points are positioned outside that range. The probable reason is the local change in the emissivity of the paint, but even with this, the standard deviation for this set is low and equals 0.102 K.

The proposed numerical method used to solve the inverse problem allows us, for instance, to determine the surface temperature and the local heat flux as seen in Fig. 7 as the function of wall superheat in FC-72 boiling on the aluminium fin at ambient pressure. For the sake of comparison, a similar relationship devised using a 1-D model is shown. It is also worth noting that in this particular range, the maximum $Bi \approx 0.08$ is found at the wall superheat ~20 K. As can be seen from the two figure s above, the good convergence of the calculation results obtained with two different methods using the same experimental data confirms that the method proposed for solving the ill-conditioned inverse problem is correct.

Similarly, the relationship between the local heat flux removed from the surface and the wall superheat is determined from the distribution of temperature measured at the outer surface of the 12 mm fin, according to the 1-D procedure. This leads to an exponential dependence of q on θ, as shown by the straight line in Fig. 7. The constant exponent can then be determined from its slope. The exponent value 1.52 indicates the presence of only nucleate boiling on the whole fin surface. Moreover, this exponent is twice as small as that given in the Rohsenow correlation [6].

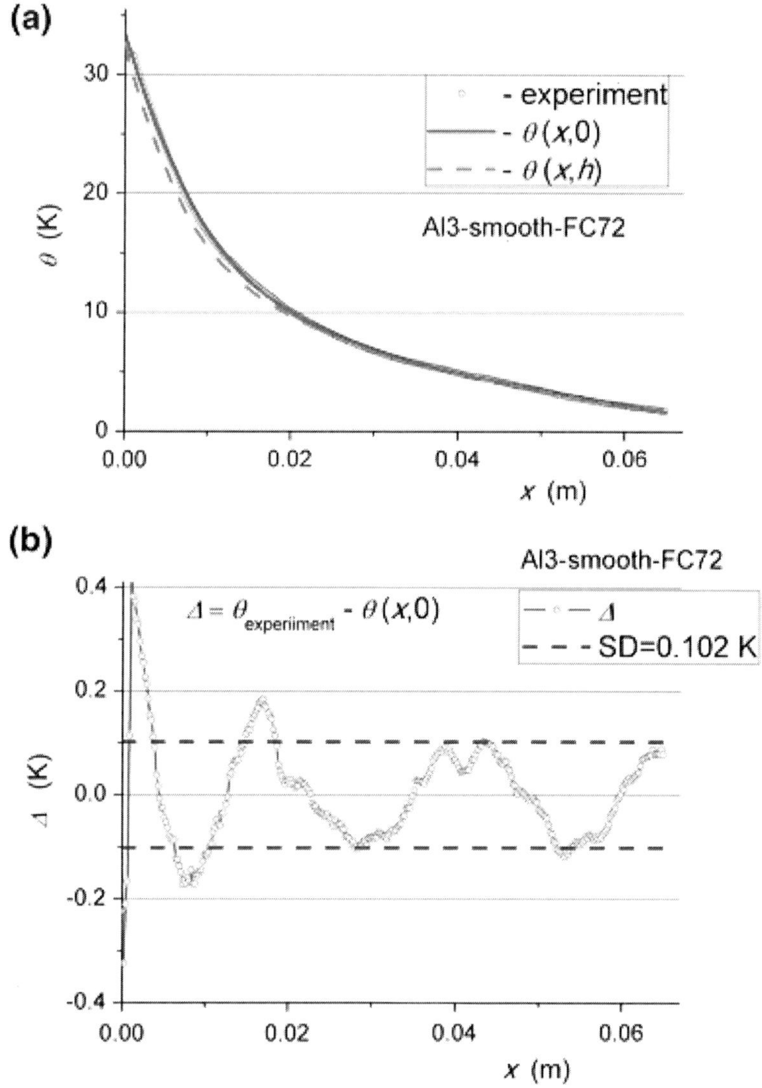

Figure 6. Results obtained for a 3 mm thick aluminium fin: **a** measured and calculated wall superheats at the outer OA and inner CB lines (see Fig. 4), *θ(x,0)* and *θ(x,h)* respectively, **b** temperature differences between the measured and calculated points at the surface viewed through a thermovision camera (y = 0)

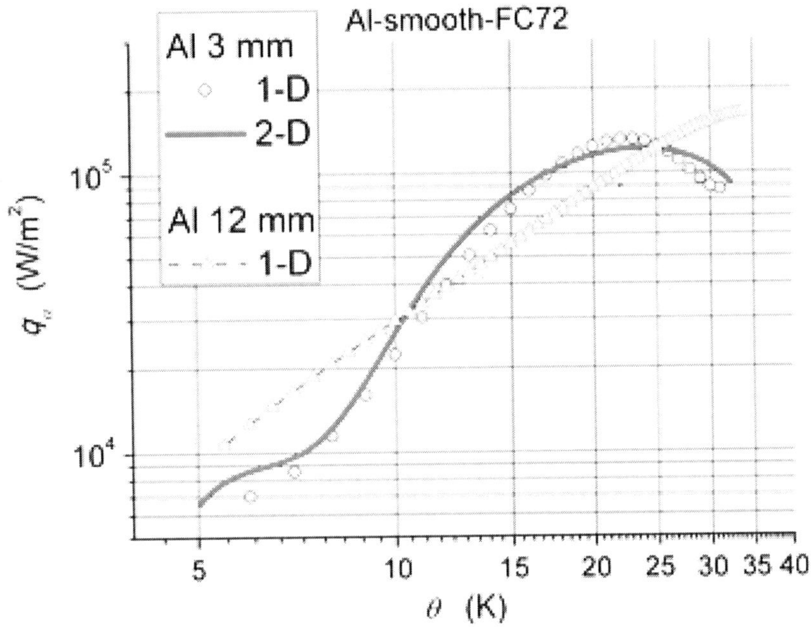

Figure 7. Measured and calculated heat flux for a thin aluminium fin

The boiling curves plotted for the two specimens in accordance with the 1-D procedure differ in shape and value. This falsely suggests that the physical phenomena present during the phase change heat transfer differ for the two specimens made of the same material and having the same surface morphology. Since this is not true, proper determination of the boiling heat transfer processes in thick fins requires using a numerical method, which, even for ill-posed problems, gives correct results.

5 DISCUSSION

The local values of heat flux density derived in Sect. 4 can be used to calculate the heat transfer coefficient for each wall superheat. As its maximum value is about 4800 W/m^{2}K, the Biot number for the 3 mm thick fin is smaller than 0.1. For a fin with the same physical parameters but a greater thickness, e.g. 12 mm, the temperature changes through the thickness cannot be neglected while determining the heat transfer conditions. The Biot number for the 12 mm thick fin is about 0.32, so the problem must be at least two-dimensional. Systems with an easy-to-observe surface can have the thermal field measured; the boiling heat transfer coefficient can be calculated using the algorithm described in Sect. 3.2. For the distribution of temperature measured at the fin surface (Fig. 3), it was essential to assume the boundary conditions (5)–(8), develop an approximation of $\Theta(x,y)$, and calculate the coefficients c_j from the best fit condition. The function determined in this way describes the thermal field in the OABC area (see Fig. 4).

Figure 8 presents the measured and calculated values of the wall superheat for two surfaces of the 12 mm thick fin, i.e. $y = 0$ and $y = h$ (see Fig. 4). By comparison, during boiling heat transfer, e.g. by natural convection or by forced convection with no phase change, large amounts of heat flux are removed. The largest are in the nucleate boiling regime. As shown in Fig. 7, for the smooth aluminium sample with a thickness of 12 mm, the process starts at a wall superheat of about 8 K at a distance of about 3 cm from the base (see Fig. 8). With a rise in the wall superheat in the nucleate boiling regime, there is an increase in the heat transfer coefficient. Heat flux removal increases locally, resulting in an increase in the difference in temperature between the outer surface (y = 0) and the inner surface (y = h). The value exceeds 5 K at points located near the base of the 12 mm thick fin. This fact cannot be ignored in the analysis of heat transfer in fins with such a geometry.

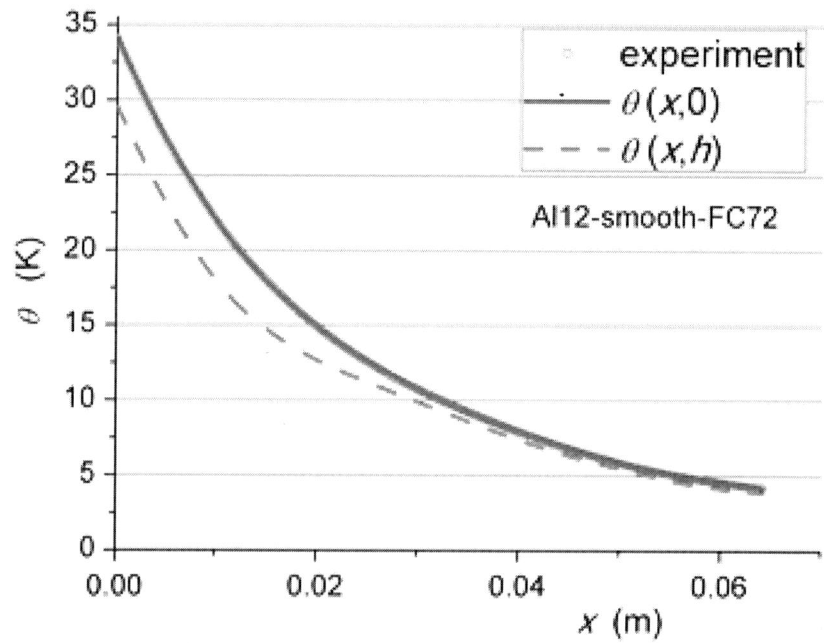

Figure 8. Measured and calculated wall superheat for a 12 mm thick aluminium fin

Figure 9 shows a gradient of the wall superheat over the outer surfaces ($y = 0$ and $y = h$) along the fin length. The calculated values result from the assumed boundary conditions. It should be noted that a good convergence of the measurement results with the assumed values is found.

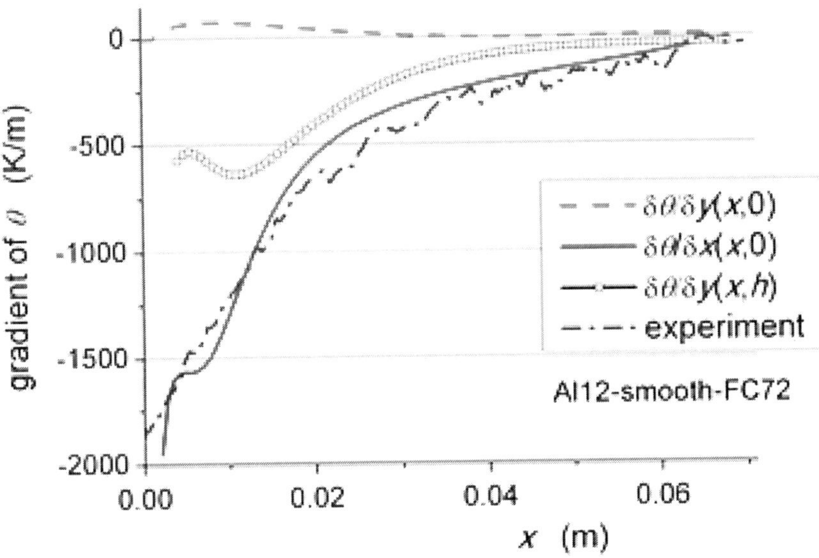

Figure 9. Temperature gradients along the fin length

However, some disagreement with Eq. 7 is observed for the temperature at the base of the fin. As can be seen from Fig. 10, the distribution of the wall superheat for $x = 0$ changes with the height. The variability of this quantity is a result of a very high heat transfer coefficient, and it seems to correct the assumed boundary condition.

Figure 10 shows the numerically calculated distribution of temperature in the 12 mm thick fin in selected cross-sections. There is a significant change in temperature in the fin cross section to the height of about 30 mm from the base. Near the surface, where nucleate boiling exists, the temperature drops considerably. Beyond this area, the decrease in temperature is insignificant.

In the numerical calculations, the approximation of the exact solution (13) is employed to determine the thermal field, which can be used to calculate the temperature gradients. In accordance with the Fourier law, this gradient, multiplied by the coefficient of heat conductivity, determines the heat flux density. On the other hand, the one calculated in the perpendicular

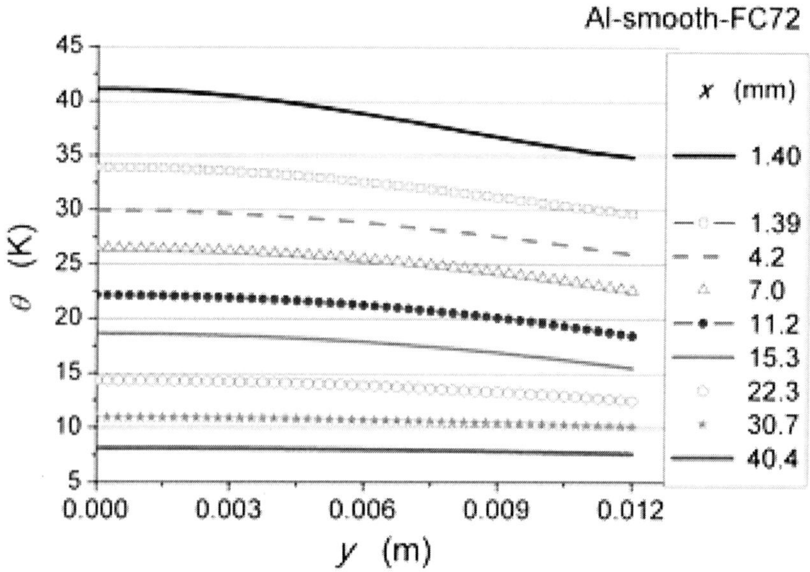

Figure 10. Temperature distribution at different cross-sections of the fin

direction on the inner edge ($y = h$) is equal to the boiling heat transfer on the fin:

$$q_\alpha = -\lambda \left. \frac{\partial \Theta}{\partial y} \right|_{y=h} \quad \text{and} \quad \alpha = \frac{q_\alpha}{\Theta} \tag{19}$$

The calculation result in the perpendicular direction on the inner edge ($y = h$) of the fin is presented in the form of a boiling curve in Fig. 11.

The boiling curve is defined locally, i.e. for the local values of heat flux and heat transfer coefficients from wall superheat. The temperatures are taken at the inner surfaces of the fins. For the 3 mm thick fin, the temperature values at the inner and outer surfaces are nearly the same. For the thicker fin, the difference is significant, as shown in Figs. 8 and 10. The total heat dissipation for the two fins is given in Figs. 2 and 3, respectively.

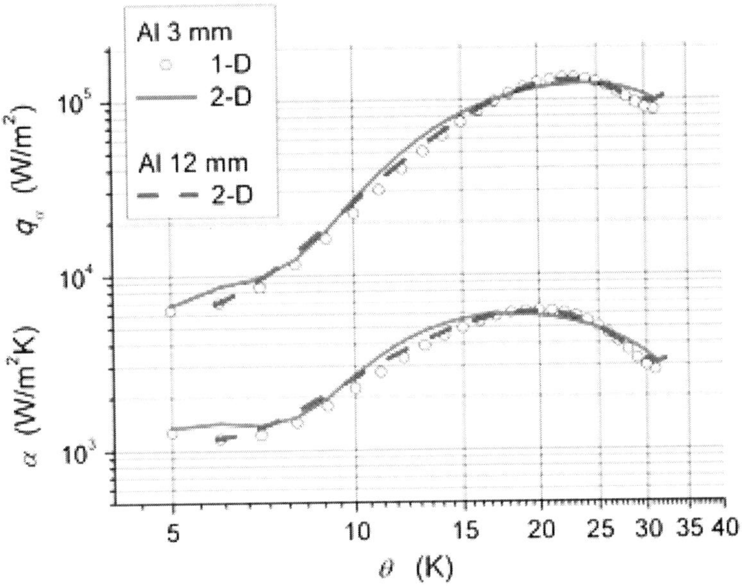

Figure 11. Boiling curve for FC 72 at ambient pressure on an aluminium fin

The analysis of heat transfer for refrigerant FC 72 boiling at ambient pressure over an aluminium fin shows that two steady boiling regimes on the surface: nucleate boiling and free convection are found. The boiling curve can be divided into three characteristic regimes. In the low superheat regime (up to about 6.5 K), free convection at a nearly constant value of the heat transfer coefficient is observed. In the range of the wall superheat higher than 7 K, two subregimes can be distinguished. When the wall superheat is in the range from about 7 K to ~8.5 K, single bubbles lift off the surface and cause more intensive mixing of the fluid close to the fin surface. When the wall superheat grows above 9 K, more nucleation sites become active and the nucleation boiling regime occurs. As a result, a rapid rise in the heat flux density and the heat transfer coefficient is observed.

Due to the very high coefficient of heat transfer in the nucleate boiling regime, in order to ensure a good representation of one-dimensional heat conduction, it is necessary to prepare the fin of a material with high thermal conductivity and low thinness. Both of those affect the value of the Biot number, which should be as small as possible. Therefore, studies were performed for aluminium fins with a thickness of 3 mm, for which at the maximum heat transfer coefficient, $Bi \approx 0.08$ and gradually decreases with drop of the local heat transfer coefficient.

Figure 6a shows the wall superheat along the height of the fin. The Figure presents measurement result and two lines for $y = 0$ and $y = h$, which were determined numerically by the given algorithm. From the comparison of the temperature difference from the side of thermovision camera shown in Fig. 6b, the errors of the method could be drawn out. From the set under consideration, it can be seen that almost all values are within the range ±0.2 K and the standard deviation equals 0.102 K.

For the fin made of aluminium with a thickness of 12 mm, temperature distribution in its cross OABC was calculated (see Fig. 4), and the results are shown in Figs. 8 and10. Temperature difference between inner and outer surfaces at the base of the fin is considerable and reaches slightly above 5.5 K. The difference of these values below 0.5 K occurs only at a distance greater than 35 mm from the base, where free convection boiling is observed.

Figure 9 shows the temperature gradients. Its value in the direction of the y-axis is almost zero, which indicates the correct adoption of the boundary condition (6). The other two curves in this Figure are the result of good agreement of measured and calculated temperatures at the surface, as discussed above.

For the commercial surface with lots of microcavities, the nucleate boiling regime starts at a wall superheat of about 9 K and, as more nucleation sites become active, and grows rapidly with the superheat increase, as shown in the boiling curve in Fig. 11. For superheat above 8 K, the heat transfer coefficient gradually increases and its maximum is about ~6.2 kW/m^2K at the wall superheat of 20 K. The increased generation of vapor causes bubble interference and coalescence, and the vapour escapes in forms of jets or columns, which subsequently merge into slugs. This is accompanied by a much slower increase in the heat transfer coefficient and at the wall superheat above ~23.5 K, a consequent decrease in the heat flux density is observed.

6 CONCLUSIONS

The aim of the experimental research was to measure the local values of the heat flux and the boiling heat transfer coefficient on a non-isothermal fin surface. The thermal conductivity of aluminium under nucleate boiling conditions is relatively high, but 12 mm thickness of the fin makes the Biot number much higher than 0.1.

The study involved comparing two fins with the same surface morphology but different thicknesses. For the thinner fin, the Biot number was smaller than 0.1, while for the thicker fin the number was much greater. The boiling curves plotted for the two specimens in accordance with the 1-D procedure described in Sect. 2.1 are clearly different from each other in shape and value (see Fig. 7). The maximum difference is more than 31 % at a wall superheat of 18 K.

Since it is impossible to use a 1-D model of heat transfer for such a system, the calculations were conducted using a two-dimensional heat conduction equation, which is an ill-conditioned problem due to the measurement method applied.

To validate the method of calculation, a series of measurements was performed, whose purpose was to estimate the local heat transfer coefficient under conditions as close as possible to one-dimensional temperature distribution and to compare those results with the analogous ones obtained using the method in the inverse problem.

The performance of 1-D fin with a constant cross-section is analysed on the basis of a one-dimensional expression for the temperature distribution. With an assumption that the outcome of integration is an holomorphic function, its value is represented by a power series. By differentiating this expression, functional dependences for the local heat fluxes and heat transfer coefficients are obtained. In this way the corresponding boiling curve for the investigated surface is given and thoroughly discussed in the paper.

The results of numerical calculations are the temperature values inside the area and on the invisible side (see Fig. 10). They are described by the formula (13). On this basis, the local heat flux density and corresponding heat transfer coefficient are evaluated, as shown in Figs. 7 and 11. There, a correlation between the values calculated assuming a one-dimensional model of heat transport phenomena in the fin is also presented. Visible reduction in heat transfer coefficient in the high heat fluxes range is the result of polynomial approximation used in the model 1-D and cannot be here the basis for inference about the location of the first critical point of boiling.

The approximation of the test results shown in Fig. 5 was performed with sixth degree polynomial. Eq. (3) was used to determine the local heat flux or the heat transfer coefficient:

$$\alpha = \frac{q_\alpha}{\theta} = a_0 + a_1\theta + a_2\theta^2 + a_3\theta^3 + a_4\theta^4 + a_5\theta^5 \quad .(20)$$

The coefficients were as follows: a_0 = 5470.05 W/m^{2}K, $a_1/a_0 = -0.30757$ K^{-1}, $a_2/a_0 = 0.034797$ K^{-2}, $a_3/a_0 = -0.00094$ K^{-3}, $a_4/a_0 = -9.3\text{E}-06$ K^{-4}, $a_5/a_0 = 4.24\text{E}-07$ K^{-5} (valid in the range of θ = 6–31 K).

To sum up, the analysis performed shows that the proposed method provides accurate results coincident with 1D ones. The obtained results of this study show that the method proposed is accurate and efficient for solving the engineering heat transfer problems where the thermal field could be measured and corresponding heat fluxes are sought.

REFERENCES

1. Dhir VK (1998) Boiling heat transfer. Annu Rev Fluid Mech 30:365–401
2. Ahn HS, Kim M, Kaviany M, Kim MH (2014) Pool boiling experiments in reduced graphene oxide colloids. Part I—boiling characteristics. Int J Heat Mass Transf 74:501–512
3. Ahn HS, Kim M, Kaviany M, Kim MH (2014) Pool boiling experiments in reduced graphene oxide colloids. Part II—behavior after the CHF, and boiling hysteresis. Int J Heat Mass Transf 78:224–231
4. Naphon P, Thongjing C (2014) Pool boiling heat transfer characteristics of refrigerant-nanoparticle mixture. Int Commun Heat Mass Transf 52:84–89 .
5. Ahn HS, Kim MH (2013) The boiling phenomenon of alumina nanofluid near critical heat flux. Int J Heat Mass Transf 62:718–728
6. Pioro IL (1999) Experimental evaluation of constants for the Rohsenow pool boiling correlation. Int J Heat Mass Transf 42:2003–2013
7. Pioro IL, Rohsenow W, Doerfer SS (2004) Nucleate pool-boiling heat transfer. II: assessment of prediction methods. Int J Heat Mass Transf 47:5045–5057 .
8. Kolev NI (1995) How accurately can we predict nucleate boiling? Exp Therm Fluid Sci 10:370–378
9. Adib TA, Vasseur J (2008) Bibliographic analysis of predicting heat transfer coefficients in boiling for applications in designing liquid food evaporators. J Food Eng 87:149–161
10. Shah MM (2005) Evaluation of general correlations for heat transfer during boiling of saturated liquids in tubes and annuli. ASME summer heat transfer conference USA, HT2005-72025

11. Orzechowski T, Tyburczyk A (2013) Heat transfer on fins operating at boiling—experimental and numerical procedure. In: EPJ web of conferences: experimental fluid mechanics 2013 pp 523–529

12. Roh H-S (2014) Heat transfer mechanisms in pool boiling. Int J Heat Mass Transf 68:332–342

13. Orzechowski T (2001) Analysis of the boiling heat transfer from a microstructure-covered fin. Arch Thermodyn 22:77–94

14. Krikkis R, Sotirchos SV, Razelos P (2004) Analysis of multiplicity phenomena in longitudinal fins under multi-boiling conditions, ASME. J Heat Transf 126:1–7

15. Liaw S-P, Yeh R–H, Yeh W–T (2005) A simple design of fins for boiling heat transfer. Int J Heat Mass Transf 48:2493–2502

16. Gonzalez-Fernandez CF, Alarcon MF, Alhama F (2004) Transient multiboiling in a pin fin with temperature dependent thermal conductivity. Heat Mass Transfer 41:67–74

17. Turkyilmazoglu M (2012) Exact solutions to heat transfer in straight fins of varying exponential shape having temperature dependent properties. Int J Therm Sci 55:69–75

18. Kundu B, Lee K-S (2012) Analytic solution for heat transfer of wet fins on account of all nonlinearity effects. Energy 41:354–367

19. Taler J (1988) A general method for the experimental determination of local transient heat transfer coefficients. Wärme- und Stoffübertragung 23:283–289

20. Chen H-T, Lai S-T, Haung L-Y (2013) Investigation of heat transfer characteristics in plate-fin heat sink. Appl Therm Eng 50:352–360

21. Torabi M, Aziz A (2012) Thermal performance and efficiency of convective–radiative T-shaped fins with temperature dependent thermal conductivity, heat transfer coefficient and surface emissivity. Int Commun Heat Mass Transf 39:1018–1029

22. Konda Reddy B, Balaji C (2012) Estimation of temperature dependent heat transfer coefficient in a vertical rectangular fin using liquid crystal thermography. Int J Heat Mass Transf 55:3686–3693

23. Díez LI, Espatolero S, Cortés C, Campo A (2010) Thermal analysis of rough micro-fins of variable cross-section by the power series method. Int J Therm Sci 49:23–35

24. Huang C-H, Chang W-L (2012) An inverse design method for optimizing design parameters of heat sink modules with encapsulated chip. Appl Therm Eng 40:216–226

25. Grysa K, Maciąg A, Pawinska A (2012) Solving nonlinear direct and inverse problems of stationary heat transfer by using Trefftz functions. Int J Heat Mass Transf 55:7336–7340

26. Das R, Mallick A, Ooi KT (2013) A fin design employing an inverse approach using simplex search method. Heat Mass Transf 49:1029–1038

27. Zueco J, Alhama F, González Fernández CF (2005) Numerical nonlinear inverse problem of determining wall heat flux. Heat Mass Transf 41:411–418

28. Orzechowski T (2007) Determining local values of the heat transfer coefficient on a fin. Exp Therm Fluid Sci 31:947–955

29. Ciałkowski M, Frąckowiak A (2000) Heat functions and their application to solving heat conduction and mechanical problems. Poznań University of Technology, Poznan **(in Polish)**

3
CHAPTER

Performance of Siloxane Mixtures in a High-Temperature Organic Rankine Cycle Considering the Heat Transfer Characteristics during Evaporation

Theresa Weith *, Florian Heberle, Markus Preißinger and Dieter Brüggemann

Lehrstuhl für Technische Thermodynamik und Transportprozesse, Zentrum für Energietechnik, Universität Bayreuth, Universitätsstraße 30, 95447 Bayreuth, Germany

ABSTRACT

The application of the Organic Rankine Cycle to high temperature heat sources is investigated on the case study of waste heat recovery from a selected biogas plant. Two different modes of operation are distinguished: pure electric power and combined heat and power generation. The siloxanes hexamethyldisiloxane (MM) and octamethyltrisiloxane (MDM) are chosen as working fluids. Moreover, the effect of using mixtures of these components is analysed. Regarding pure electricity generation, process simulations using the simulation tool Aspen Plus show an increase in second law efficiency of 1.3% in case of 97/03 wt % MM/MDM-mixture, whereas for the combined heat and power mode a 60/40 wt % MM/MDM-mixture yields the highest efficiency with an increase of nearly 3% compared to most efficient pure fluid. Next to thermodynamic analysis, measurements of heat transfer coefficients of these siloxanes as well as their mixtures are conducted and Kandlikar's correlation is chosen to describe the results. Based on that, heat exchanger areas for preheater and evaporator are calculated in order to check whether the poorer heat transfer characteristics of mixtures devalue their efficiency benefit due to increased heat transfer areas. Results show higher heat transfer areas of 0.9% and 14%, respectively, compared to MM.

Keywords: Organic Rankine Cycle; siloxanes; high temperature heat source; zeotropic mixtures; heat transfer coefficients; heat exchange area

1. INTRODUCTION

The use of zeotropic mixtures in Organic Rankine Cycle (ORC) applications has been widely investigated during the last years. The focus of this research is mainly on low temperature applications like geothermal systems [1,2,3,4]. In these cases an efficiency increase of up to 20% is possible by using mixtures compared to the most efficient pure component due to a better glide matching during condensation and evaporation caused by non-isothermal phase change of zeotropic mixtures. Recent investigations also Figure the combination of zeotropic mixtures and a transcritical mode of operation due to a better matching of temperature profiles [5,6]. Next to the non-isothermal phase change during condensation resulting from the use of zeotropic mixtures, the evaporation phase change also occurs non-isothermally, due to the supercritical mode of operation. This results in an increase in second law efficiency of up to 60% [5]. Beside geothermal power generation, the benefit of using zeotropic mixtures is also investigated for other specific applications, like for example in the field of heat recovery from engines [7,8]. Furthermore, the first studies that focus on zeotropic mixtures for high temperature ORC have been published. Along with Angelino and Colonna [9,10], who investigated multi-component mixtures of linear siloxanes, Dong *et al.* [11] show results of the performance of hexamethyldisiloxane (MM)/octamethyltrisiloxane (MDM)-mixtures at a heat source temperature of 280 °C. Compared to low temperature applications, the efficiency increase achieved with mixtures in high temperature ORCs is less. Chys *et al.* [12] describe a benefit of nearly 16% for a heat source temperature of 150 °C, whereas at 250 °C the increase is less than 6%.

Next to this advantage of zeotropic mixtures with respect to cycle efficiency, there is also an inherent drawback of using them in ORC-applications caused by their generally reduced heat transfer characteristics compared to pure fluids [13,14]. This fact is often neglected when analysing mixtures in ORC-systems, but regarding sizing of heat exchangers the knowledge of this degradation in heat transfer coefficient is important. Looking at common refrigerants, measurements of heat transfer coefficients of those as well as their mixtures are already available in literature as their long-serving use in refrigeration-systems has driven this research [15,16,17,18,19,20]. However, in case of siloxane mixtures as well as pure siloxanes up to now there has been a lack of data.

The present study shows a thermodynamic analysis of a high temperature ORC using the linear siloxanes hexamethyldisiloxane (MM) and octamethyltrisiloxane (MDM) as well as their mixtures. Moreover, the results of measurements of heat transfer coefficients of these fluids and their mixtures in steps of 20 wt % MM are presented. Kandlikar's correlation is adapted to the experimental values and is then used for calculating the required heat exchanger area for preheating and evaporation.

2. THERMODYNAMIC PROCESS ANALYSIS

In this study the recovery of waste heat from biogas plants is chosen exemplarily as one application of the ORC. The specifications of the exhaust gas from the biogas-engine (MAN E 2842 LE 322, MAN Truck & Bus AG, Nuremberg, Germany) of the selected biogas plant are summarized in Table 1.

Table 1. Parameters of heat source (EG).

$T_{EG\text{-}IN}$ (K)	$\dot{m}_{EG}$ (kg/s)	p_{EG} (bar)	Composition (mol %)				
			N_2	H_2O	CO_2	O_2	Ar
733.15	0.48	1	64.55	18.15	9.74	6.63	0.93

Siloxanes were chosen as a working fluid due to their good performance in high temperature applications [9,21,22]. In particular, the two linear siloxanes hexamethyldisiloxane (MM) and octamethyltrisiloxane (MDM) as well as their mixtures are investigated. First degradation measurements of MM show its thermal stability up to a temperature of 300 °C [23]. Therefore, with regard to the high heat source temperature, an additional intermediate circuit is considered in the simulations in order to prevent the working fluid from decomposition. Dowtherm G is chosen as thermal oil due to its high thermal stability [24].

2.1. Methodology

Simulations have been performed using the simulation tool Aspen Plus [25]. As property method the PENG-ROB (Peng-Robinson equation of state) model is used for the exhaust gas, air, and Dowtherm G as well as the ORC-fluid. In the case of water, STEAM-NBS (NBS/NRC Steam Tables) is applied [26]. Figure 1 shows the flowsheet of the process. It consists of the ORC, in which an internal recuperator is applied for the purpose of efficiency improvement, and the additional thermal oil loop. All heat exchangers are operating in counter flow mode and pressure drops are neglected.

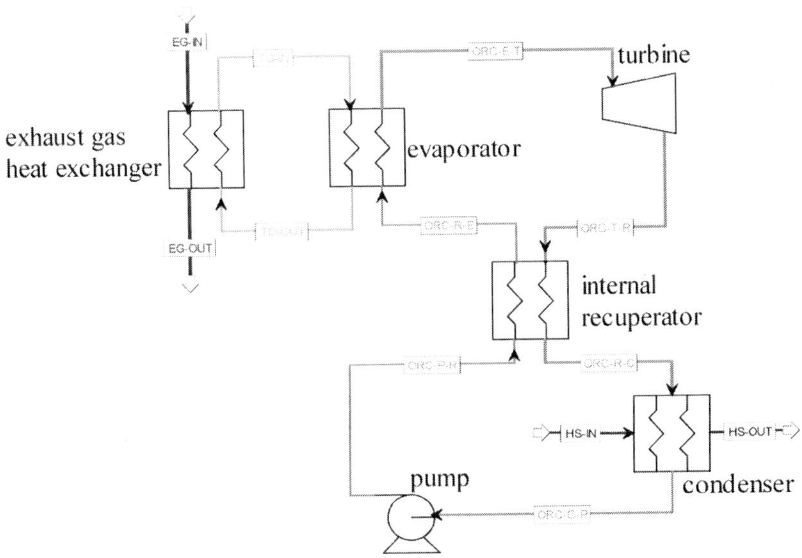

Figure 1. Aspen Plus flowsheet.

For the heat sink, two different scenarios are analysed. The first one represents pure electricity generation out of waste heat assuming direct air cooled condensation, whereas the second one considers an additional use of the condenser duty in a heating network using water. Table 2 lists the general parameters of the heat sink for those two cases.

Table 2. Heat sink (HS) parameters.

Case	Fluid	$T_{HS\text{-}IN}$ (K)	$T_{HS\text{-}OUT}$ (K)	$\dot{m}_{HS}$ (kg/s)	p_{HS} (bar)
heat sink (electricity)	air	288.15	303.15	variable	1
heat sink (CHP)	water	323.15	343.15	variable	2

As boundary conditions, the pinch point differences in the heat exchangers are kept constant (Table 3). Only the pinch point differences in the condenser and internal recuperator can reach values higher than those given in Table 3. In the condenser this occurs due to the limitation of the lower working pressure to a minimum value of 0.1 bar. Regarding fluid mixtures, partial condensation can take place in the internal recuperator. If this occurs, the boundary condition is switched. Instead of the fixed pinch point temperature difference a vapour fraction of one is set for the hot outlet stream of the recuperator (ORC-R-C). As a result, pinch point differences can reach higher values. Due to the heat source parameters, the investigated systems are small-scale ORCs with an electrical power output in the kilowatt range. Therefore, the isentropic efficiency of the turbine is set to the relatively small value of 60%.

Table 3. Boundary conditions.

$PP_{\text{EG heat exchanger}}$ (K)	$PP_{\text{evaporator}}$ (K)	$PP_{\text{condenser}}$ (K) (min)	$PP_{\text{internal recuoerator}}$ (K) (min)	η_{turbine} (%)	$\eta_{\text{generator}}$ (%)	η_{pump} (%)	η_{driver} (%)
20	30	10	10	60	90	70	90

To compare the simulation results of the different fluids, the exergetic efficiency is chosen as a criterion. For pure generation of electric energy, the exergetic efficiency is defined by the net power P_{net} (power of turbine $P_{\text{el,turb}}$ minus power of pump P_{pump}) divided by the exergy of heat source $\dot{E}_{\text{EG}}$:

$$\eta_{\text{ex,electricity}} = \frac{P_{\text{net}}}{\dot{E}_{\text{EG}}} = \frac{P_{\text{el,turb}} - P_{\text{pump}}}{\dot{m}_{\text{EG}}\left[h_{\text{EG,in}} - h_0 - T_0\left(s_{\text{EG,in}} - s_0\right)\right]} \quad (1)$$

For calculating $\dot{E}_{\text{EG}}$, mass flow rate $\dot{m}_{\text{EG}}$, enthalpy $h_{\text{EG,in}}$ as well as entropy of heat source $s_{\text{EG,in}}$ are used. The index 0 corresponds to the reference state (288.15 K, 1 bar). In the case of combined heat and power generation the exergy content of the condenser duty $\dot{E}_{\text{cond}}$ has to be taken into account:

$$\eta_{\text{ex, CHP}} = \frac{P_{\text{net}} + \dot{E}_{\text{cond}}}{\dot{E}_{\text{EG}}} \quad (2)$$

$$\dot{E}_{\text{cond}} = \dot{m}_{\text{HS}}\left[h_{\text{HS,out}} - h_{\text{HS,in}} - T_0\left(s_{\text{HS,out}} - s_{\text{HS,in}}\right)\right] \quad (3)$$

It is calculated by using inlet and outlet enthalpies ($h_{HS,in/out}$) and entropies ($s_{HS,in/out}$) of the cooling water as well as its mass flow rate $\dot{m}_{HS}$.

For evaluating the performance during evaporation and condensation, the irreversibilities of the evaporator and condenser are calculated by using respective inlet and outlet enthalpies and entropies:

$$\dot{I}_{evap} = T_0 \dot{m}_{ORC} \left[s_{ORC\text{-}E\text{-}T} - s_{ORC\text{-}R\text{-}E} - \frac{h_{ORC\text{-}E\text{-}T} - h_{ORC\text{-}R\text{-}E}}{T_{m,evap}} \right] \quad (4)$$

$$\dot{I}_{cond} = T_0 \dot{m}_{ORC} \left[s_{ORC\text{-}C\text{-}P} - s_{ORC\text{-}R\text{-}C} - \frac{h_{ORC\text{-}C\text{-}P} - h_{ORC\text{-}R\text{-}C}}{T_{m,cond}} \right] \quad (5)$$

The logarithmic mean temperatures are defined as follows:

$$T_m = \frac{T_{in} - T_{out}}{\ln\left(\frac{T_{in}}{T_{out}}\right)} \quad (6)$$

In case of the condenser T_{in} and T_{out} describe the inlet and outlet temperatures of the heat sink. For the irreversibilities of the evaporator the temperatures of Dowtherm G have to be used.

2.2. Results of Process Simulation

Due to the available heat source, the applied ORC shows power outputs in the low power range ($P_{net} < 30$ kW). The results of the process simulation are displayed in Figure 2. On the left hand side the exergetic efficiency against the mass fraction of MM in case of pure electricity generation is presented, whereas the results for the combined heat and power case are shown on the right hand side. Next to the efficiency, the corresponding upper working pressure, that yields the highest efficiency in each case, is added in the diagrams.

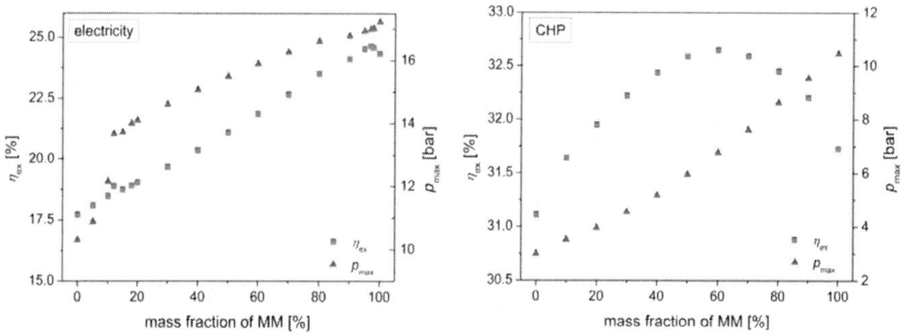

(a) (b)

Figure 2. Exergetic efficiencies and respective maximum working pressures against mass fraction of MM ((**a**): pure electricity generation, (**b**): combined heat and power generation).

Regarding the case of pure electricity generation and using steps of mass fraction of MM of 10%, it seems that there is a nearly linear behaviour of exergetic efficiency with respect to mass fraction of MM. A closer look reveals that at high mass fractions of MM there might be a deviation from this trend. Therefore, additional steps of 95%, 97% and 98% MM are added and an efficiency increase of nearly 1.3% compared to the most efficient pure component, MM, occurs for 97% MM. A similar effect between mass fractions of 10% and 20% occur, but efficiencies do not exceed that of MM.

A quite different trend with respect to mass fraction of MM can be seen for the combined heat and power production. Compared to the pure fluid with the best exergetic efficiency (MM: η_{ex} = 31.73%) an efficiency increase of 2.9% is possible when using mixtures (60 wt % MM: η_{ex} = 32.65%). Next to this improved cycle efficiency, lower working pressures are needed compared to pure MM. As shown in Figure 3, this increase in second law efficiency derives from an irreversibility minimum in the condenser. Regarding the irreversibilities in the evaporator, smaller differences between the fluids can be observed. Therefore, they do not affect the overall cycle efficiency.

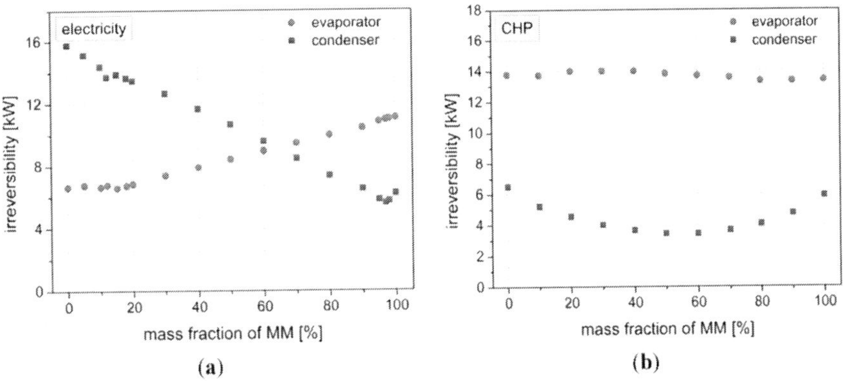

Figure 3. Irreversibilities in evaporator and condenser in case of (**a**) pure electricity generation and (**b**) combined heat and power generation.

The reason for this behaviour can be revealed by regarding the temperature profiles during evaporation and condensation (Figure 4). It can be seen that the temperature glide during evaporation is negligible compared to that during condensation. Therefore, as also published by other authors [3], a good glide matching during condensation, which results in a lower average heat rejection temperature of the cycle, is decisive for cycle efficiency. Figure 4b fairly shows the accordance in the slope of the temperature profile of heat sink and the most efficient mixture.

All in all, the results show that an increase in efficiency of nearly 1.3% (electricity) or 3% (CHP) is possible by using mixtures of MM and MDM, but, whether a use of mixtures is really advantageous or not depends, next to the pure thermodynamic performance, on several additional aspects. In this regard, the effect of an eventual composition shift on the cycle efficiency has to be considered. Moreover, the influence of mixtures on the performance of cycle components (e.g., the respective turbine efficiency) has to be evaluated. As heat transfer coefficients of mixtures are said to be lower than those of the pure components, it is crucial to investigate the required size of heat exchangers. In order to estimate this heat transfer area, proper correlations to describe the heat transfer characteristics are necessary. To identify them, measurements of heat transfer coefficients have to be conducted. The main parameter for a waste heat recovery system is the coupling with the heat source and, therefore, the evaporation process. Hence, this study focuses in a first step on evaporation heat transfer.

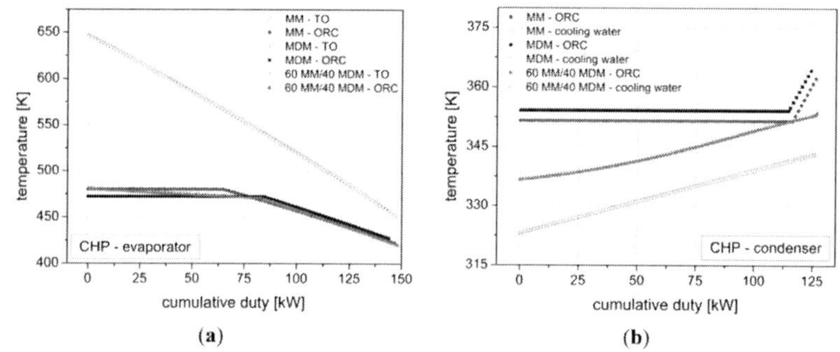

Figure 4. Temperature profiles during (**a**) evaporation and (**b**) condensation for the pure fluids as well as for the most efficient mixture (60 wt % MM) in case of CHP.

3. HEAT TRANSFER COEFFICIENTS OF SILOXANES AND SILOXANE MIXTURES

3.1. Experimental Setup

To measure the flow boiling heat transfer coefficients of the investigated siloxanes MM and MDM as well as their mixtures, a test rig (Figure 5) was built consisting of a pump, which raises the working fluid to the desired pressure, an electrically heated flow heater, which preheats the siloxane up to its saturation temperature, a test section, in which the fluid is evaporated, two expansions devices to reduce the pressure and a double-pipe heat exchanger to condense the fluid.

All components and pipes are thermally isolated to the environment by a glass wool layer. The main test section (Figure 6) consists in a smooth, horizontal tube with one U-turn, which is electrically heated by Joule effect. It is electrically isolated from the rest of the test rig by a high-temperature plastic.

Temperature and pressure is measured at the inlet as well as the outlet of the evaporator. Eleven measurement sites are placed alongside the evaporator in order to investigate heat transfer over the whole test section. At each measurement site three thermocouples are mounted at the tube. The overall outer wall temperature $T_{w,o}$ for one measurement site is therefore calculated by:

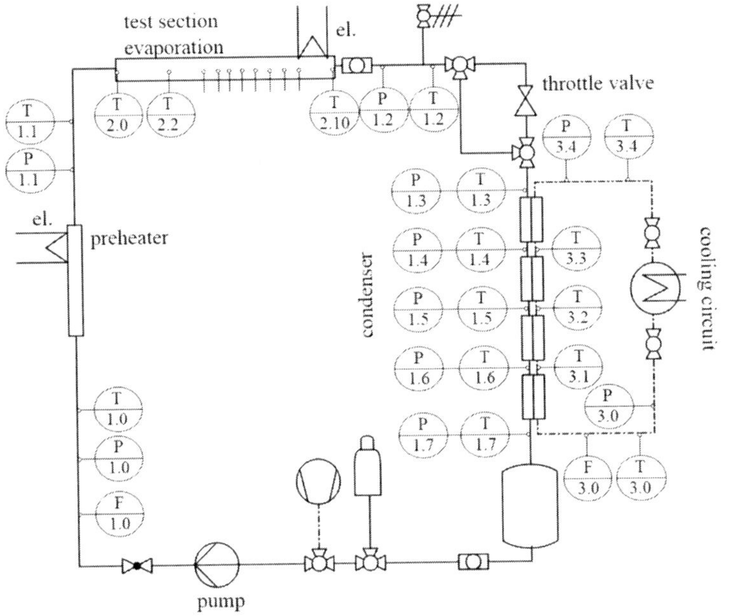

Figure 5. Test rig for measuring heat transfer coefficients.

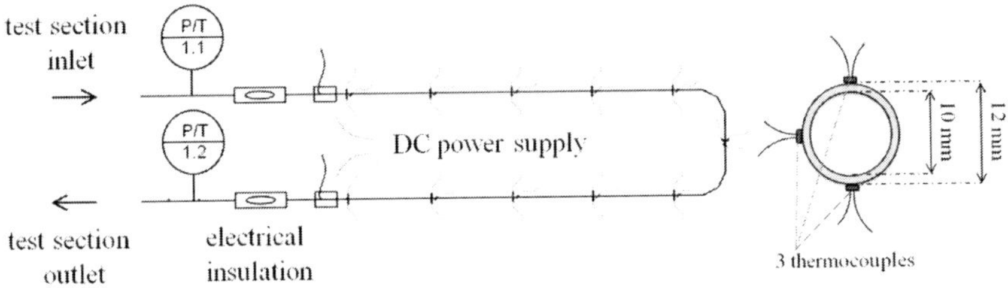

Figure 6. Test section for flow boiling.

$$T_{w,o} = \frac{T_{w,t} + 2 \cdot T_{w,m} + T_{w,b}}{4} \tag{7}$$

using the temperatures at the top ($T_{w,t}$), middle ($T_{w,m}$) and bottom ($T_{w,b}$) of the tube.

The local heat transfer coefficient of evaporation h_{evap} is evaluated by:

$$h_{\text{evap}} = \frac{\dot{q}}{T_{\text{w,i}} - T_{\text{sat}}(x,p)} = \frac{\frac{P_{\text{el}}}{A}}{T_{\text{w,i}} - T_{\text{sat}}(x,p)} \tag{8}$$

Thereby, P_{el} corresponds to the power supplied by the DC power device. The temperature at the inner face of the tube, $T_{w,i}$, can be determined with respect to the outer wall temperature, $T_{w,o}$, by applying the law of heat conduction. T_{sat} represents the bulk temperature and therefore corresponds to the respective boiling point temperature, which depends on pressure and regarding mixtures as well on vapour quality. Its value is obtained by REFPROP (Reference Fluid Thermodynamic and Transport Properties Database) [27] assuming a linear pressure profile along the test section.

3.2. Experimental Results

Experiments have been performed at constant critical pressure as well as constant saturation temperature. Mixtures in steps of 20 wt % MM have been investigated. Figure 7 shows the results of the average heat transfer coefficients during evaporation with respect to mass fraction of MM. Respective uncertainties in measurements are below 30% [28].

It is obvious that the heat transfer coefficients of the siloxane mixtures deviate from the theoretical ideal values. The ideal curve is just a linear interpolation between the heat transfer coefficients of the pure components. In fact, it does not have any real physical meaning, but it can be used to describe the effect that the heat transfer coefficients of mixtures can be lower than those of the pure components. In case of constant reduced pressure, the maximum deviation of 46% occurs at 20% MM. The heat transfer coefficient at constant saturation temperature deviates maximum 42% from the ideal value, again at 20% MM.

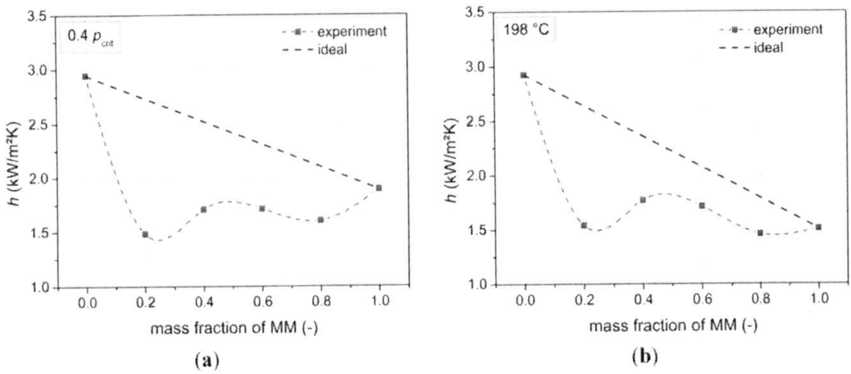

Figure 7. Average heat transfer coefficients for different mass fractions of MM at a pressure of **(a)** 0.4 p_{crit} [28] as well as **(b)** a saturation temperature of 198 °C.

3.3. Examination of Appropriate Correlations

A lot of different correlations for flow boiling of pure fluids as well as mixtures have been proposed during the last decades. Nevertheless, none of those correlations are based on high-temperature fluids like the investigated siloxanes. Due to that, as a basis, a correlation has to be chosen that offers a fluid-specific parameter in order to adjust it to the experimental data. Such correlations are given by Kandlikar [29] for pure fluids as well as mixtures. The investigation of the adaption of Kandlikar's correlations to the experiments with siloxanes and their mixtures revealed that they are able to represent the general trend of heat transfer coefficient with respect to vapour quality and mass flux density quite well [28]. Therefore, the present study focuses on this correlation.

Regarding flow boiling in plain tubes, Kandlikar [29] suggests the following expression for the heat transfer coefficient of pure fluids (α_{TP}):

$$\alpha_{TP} = \max\left(\alpha_{NBD}, \alpha_{CBD}\right) \tag{9}$$

where α_{NBD} describes the heat transfer coefficient in the nucleate boiling region:

$$\alpha_{\mathrm{NBD}} = 0.6683\,\mathrm{Co}^{-0.2}(1-x)^{0.8}\,\alpha_{\mathrm{lo}} + 1058.0\,\mathrm{Bo}^{0.7}(1-x)^{0.8}F_{\mathrm{Fl}}\alpha_{\mathrm{lo}} \tag{10}$$

whereas α_{CBD} is used for regions where convective boiling is dominant:

$$\alpha_{\mathrm{CBD}} = 1.136\,\mathrm{Co}^{-0.9}(1-x)^{0.8}\,\alpha_{\mathrm{lo}} + 667.2\,\mathrm{Bo}^{0.7}(1-x)^{0.8}F_{\mathrm{Fl}}\alpha_{\mathrm{lo}} \tag{11}$$

In both cases the heat transfer coefficient with respect to vapour quality x depends on the convection number Co, the boiling number Bo as well as on a fluid-surface parameter F_{Fl}. For the single-phase heat transfer coefficient α_{lo} the correlations by Petukhov and Popov (Equation (12), for $0.5 \le Pr_L \le 2000$ and $10^4 \le Re_{LO} \le 5 \times 10^6$) or Gnielinski (Equation (13), for $0.5 \le Pr_L \le 2000$ and $2300 \le Re_{LO} \le 10^4$) are applied:

$$\alpha_{\mathrm{lo,PaP}} = \frac{\lambda_{\mathrm{l}}}{d_{\mathrm{i}}} \frac{\mathrm{Re}_{\mathrm{lo}}\,\mathrm{Pr}_{\mathrm{L}}\,(f/2)}{1.07 + 12.7\left(\mathrm{Pr}_{\mathrm{L}}^{2/3} - 1\right)(f/2)^{0.5}} \tag{12}$$

$$\alpha_{lo,G} = \frac{\lambda_l}{d_i} \frac{(\mathrm{Re}_{\mathrm{lo}} - 1000)\,\mathrm{Pr}_{\mathrm{L}}\,(f/2)}{1.07 + 12.7\left(\mathrm{Pr}_{\mathrm{L}}^{2/3} - 1\right)(f/2)^{0.5}} \tag{13}$$

$$f = [1.58\ln(\mathrm{Re}_{\mathrm{lo}}) - 3.28]^{-2} \tag{14}$$

These correlations are based on the Reynolds number Re, the Prandtl number Pr and the friction factor f. In addition, the liquid thermal conductivity λ_l as well as the inner diameter d_i of the tube contribute to the heat transfer coefficient.

In case of fluid mixtures, Kandlikar distinguishes between three different regions according to the volatility parameter $V_1$and the boiling number Bo [29]:

1. *Near azeotropic region*—$V_1 < 0.03$. In this domain the correlation for pure fluids is adopted (Equations (9)–(14)).

2. *Moderate diffusion-induced suppression region*—$0.03 < V_1 < 0.2$ and $\mathrm{Bo} > 10^{-4}$. Heat transfer is dominated by convection. Therefore, Equation (12) is used to predict heat transfer coefficients.

3. *Severe diffusion-induced suppression region*—$0.03 < V_1 < 0.2$ and $\mathrm{Bo} \leq 10^{-4}$ or $V_1 \geq 0.2$. This region is still dominated by convection. Moreover, additional mass diffusion resistance due to large composition differences has to be taken into account. Due to that, Equation (12) is extended by the diffusion-induced suppression factor F_D:

$$\alpha_{\mathrm{CBD}} = 1.136\,\mathrm{Co}^{-0.9}(1-x)^{0.8}\,\alpha_{\mathrm{lo}} + 667.2\,\mathrm{Bo}^{0.7}(1-x)^{0.8}F_{\mathrm{Fl}}\alpha_{\mathrm{lo}}F_{\mathrm{D}} \quad (15)$$

$$F_D = \frac{0.678}{1+V_1} \quad (16)$$

where the volatility parameter V_1 is defined by:

$$V_1 = \frac{c_{p,\mathrm{l}}}{\Delta h_{\mathrm{LG}}}\left(\frac{\kappa}{D_{12}}\right)^{0.5}\left|(y_1 - x_1)\left(\frac{dT}{dx_1}\right)\right| \quad (17)$$

using the liquid specific heat $c_{p,\mathrm{l}}$, the latent heat of vaporisation Δh_{LG}, the thermal diffusivity κ and the diffusion coefficient D_{12}. Moreover, the mass fraction of component 1 in vapour (y_1) and liquid (x_1) phase as well as the slope of the bubble point curve (dT/dx_1) contribute.

At first, Kandlikar's correlation was applied to the experimental test conditions while using $F_{\mathrm{Fl}} = 1$ as suggested by Kandlikar for all fluids in stainless steel tubes. The results are demonstrated in Figure 8.

It can be seen that the absolute values between the correlation and the measurements differ, especially for the results of MM. For constant reduced pressure, the value of the experimental results of MM lies 46% below those of the correlation. In case of constant evaporation temperature the deviation is even 58%.

In order to obtain better agreement between the experimental and the theoretical heat transfer coefficients, the fluid-surface parameter F_{Fl} is used to adapt the values. It is fitted to the average heat transfer coefficients of the pure siloxanes.

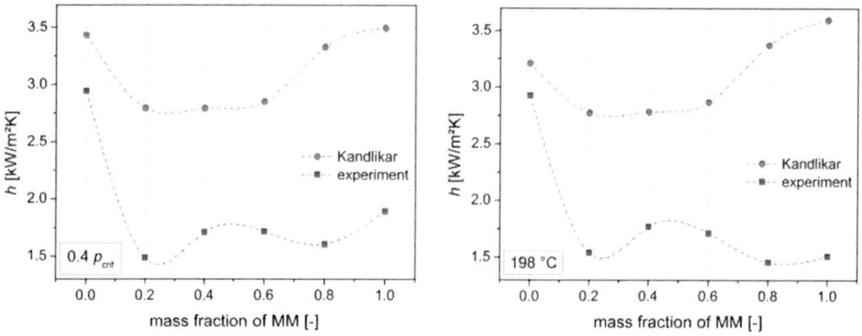

(a) (b)

Figure 8. Comparison of experimental results and Kandlikar's correlation (F_{Fl} = 1): Average heat transfer coefficients for different mass fractions of MM at a pressure of (**a**) 0.4 p_{crit} as well as (**b**) a saturation temperature of 198 °C.

Using $F_{Fl,MM}$ = 0.1 and $F_{Fl,MDM}$ = 0.8 results in maximum deviations of 15% in case of MM and 1.5% regarding MDM. For mixtures, the fluid-surface parameter is derived from those of the pure components as suggested by Kandlikar:

$$F_{\text{Fl, mixture}} = x_{\text{MM}} F_{\text{Fl,MM}} + x_{\text{MDM}} F_{\text{Fl,MDM}} \tag{18}$$

Hereby, x_{MM} corresponds to the mass fraction of MM in liquid phase, x_{MDM} to that of MDM, respectively.

The results of applying these parameter values in the correlation are presented in Figure 9. The degradation in heat transfer coefficient in case of mixtures is small compared to the experimental data. In both cases, the mean deviations of the experimental data to the correlation are below 17%. Maximum deviations appear at 20% MM with a value of nearly −37%.

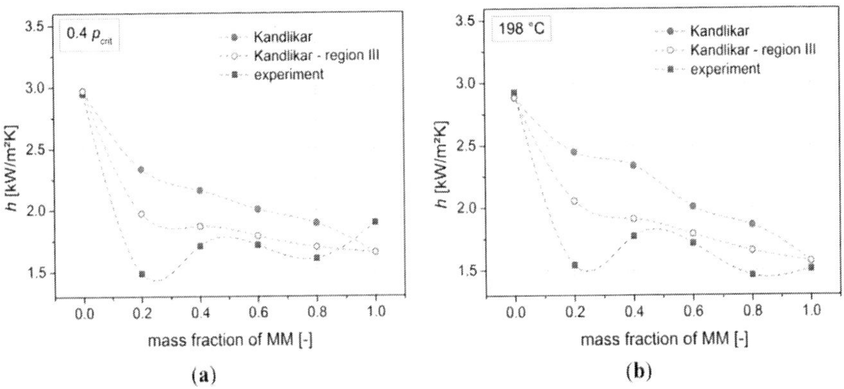

Figure 9. Comparison of experimental results and Kandlikar's correlation ($F_{FI,MM}$ = 0.1, $F_{FI,MDM}$ = 0.8): Average heat transfer coefficients for different mass fractions of MM at a pressure of (**a**) 0.4 p_{crit} as well as (**b**) a saturation temperature of 198 °C.

Due to the investigated boundary conditions, only region I and II from Kandlikar's correlation are applied. As the experimental data still deviate from the adjusted correlation, the results by using the severe diffusion-induced suppression region (region III) (even though the volatility parameter or the Bo-Number would suggest another region) are analysed, too. The respective average heat transfer coefficients are added in Figure 9. By using this equation, maximum deviations of solely 25% are present. Mean deviations of the experimental data to the correlation are below 10%. Therefore, it is recommended to use the correlation given in region III together with the adjusted values of F_{FI} when calculating heat transfer coefficients of MM/MDM-mixtures.

4. ESTIMATION OF REQUIRED HEAT EXCHANGER AREA

In order to reveal, whether the reduced heat transfer characteristics of fluid mixtures compensate their slight predominance in case of system efficiency, required heat exchange areas are investigated. For that purpose, a shell-and-tube heat exchanger with one shell pass and two tube passes is used as preheater and evaporator. Table 4 lists the main specifications and assumptions.

Table 4. Specifications and assumptions for the shell and tube heat exchanger.

Geometry Data			Maximum Velocities (m/s) [30,31]		
outer tube radius r_a (mm)	wall thickness Δd (mm)	thermal conductivity λ_{steel} (W/mK)	ORC liquid	ORC vapour	Dowtherm
9.5	2.1	15	4	20	4

The general approach for calculating the heat exchanger area is derived from Roetzel and Spang [32]. It is based on the Number of Transfer Units (NTU)-method. The NTU of the real shell and tube heat exchanger is obtained by applying a correction factor to the NTU of an ideal counterflow heat exchanger. For calculating the heat transfer coefficients at the inner and outer tube different models are applied. For Dowtherm G as well as for the ORC-fluid in the preheater, the correlation for turbulent flow by Gnielinski [33] is chosen, whereas for the evaporation of the ORC-fluid Kandlikar's correlation, as presented in Section 3, is used. The whole calculation is performed in Matlab [34]. Fluid properties are derived from REFPROP [27].

The calculations of required heat transfer area are performed for the pure fluids as well as for one more efficient mixture for the case of pure electricity and combined heat and power generation, respectively. Table 5 and Table 6 list the respective process parameters that derived from the thermodynamic analysis in Section 2. These values represent the input parameters of the heat exchanger calculations.

Table 5. Process parameters for pure electric energy generation.

Fluid	$\dot{m}_{TO}$ (kg/s)	$T_{TO\text{-}IN}$ (K)	$T_{TO\text{-}OUT}$ (K)	$\dot{m}_{ORC}$ (kg/s)	$T_{ORC\text{-}R\text{-}E}$ (K)	$T_{ORC\text{-}E\text{-}T}$ (K)	p_{ORC} (bar)
MM	0.35	645.94	449.68	0.50	419.68	511.59	17.23
MDM	0.35	645.41	502.74	0.51	472.74	541.93	10.30
MM/MDM 95/05	0.35	646.32	452.51	0.50	422.51	513.32	16.92

Table 6. Process parameters for combined heat and power generation.

Fluid	$\dot{m}_{TO}$ (kg/s)	$T_{TO\text{-}IN}$ (K)	$T_{TO\text{-}OUT}$ (K)	$\dot{m}_{ORC}$ (kg/s)	$T_{ORC\text{-}R\text{-}E}$ (K)	$T_{ORC\text{-}E\text{-}T}$ (K)	p_{ORC} (bar)
MM	0.35	646.32	452.58	0.57	422.56	480.22	10.46
MDM	0.35	647.04	457.78	0.65	427.78	472.37	3.00
MM/MDM 60/40	0.35	645.99	450.09	0.57	420.30	480.90	6.77

The results of the heat exchanger analysis are summarised in Table 7 and Table 8. In both cases MM requires the smallest heat transfer area.

For pure electricity generation MDM shows the biggest heat exchanger areas. Regarding the evaporator (*EVAP*), 19% more heat compared to MM has to be transferred, but the heat transfer coefficient of evaporation exceeds that of MM by 53%. Due to that, an additional aspect has to contribute to the increased heat transfer area of 56%. Table 5 shows that mass flow rates of all three fluids are in the same range, but MDM shows the lowest pressure as well as temperature differences. Therefore, process parameters like temperatures and the respective fluid properties affect the required area just as well as pure heat transfer characteristics of fluids.

Table 7. Results of heat exchanger calculations for the case of pure electric power generation.

Pure Electricity	Heat Exchange Area (m^2)		Number of Tubes		Heat Transfer Coefficient ($kW/m^2 \cdot K$) ORC/Dowtherm		Heat Duty (kW)	
	PH	*EVAP*	*PH*	*EVAP*	*PH*	*EVAP*	*PH*	*EVAP*
MM	0.725	0.120	2	2	5.761/7.202	26.414/8.343	113.35	30.725
MDM	0.741	0.187	2	2	5.141/7.450	40.502/8.256	86.901	36.500
MM/MDM 95/05	0.728	0.125	2	2	5.699/7.201	32.575/8.336	111.6	32.135

Table 8. Results of heat exchanger calculations for the case of combined heat and power generation.

CHP	Heat Exchange Area (m^2)		Number of Tubes		Heat Transfer Coefficient ($kW/m^2 \cdot K$) ORC/Dowtherm		Heat Duty (kW)	
	PH	*EVAP*	*PH*	*EVAP*	*PH*	*EVAP*	*PH*	*EVAP*
MM	0.583	0.270	2	3	5.756/6.942	34.394/8.660	77.76	65.276
MDM	0.536	0.379	2	9	4.869/6.806	60.753/10.312	63.873	83.795
MM/MDM 60/40	0.602	0.373	2	4	5.245/6.762	37.859/8.914	64.009	80.192

In case of the mixture the evaporator needs a 4.2% enlarged area compared to that of MM. Process parameters are quite similar. The heat transfer coefficient of the mixture lies 23.3% above that of MM, whereas a higher heat duty of 4.6% is transferred. Together with slightly reduced temperature differences a higher area is required. Regarding the whole process of heat input using the mixture, 0.9% higher heat transfer areas result compared to MM.

For the case of the combined heat and power generation similar trends occur. Regarding the evaporator, MDM requires a 40.4% higher area than MM. Heat transfer coefficient is 76.6%, heat duty 28.4% above the values of MM. A 14% higher mass flow rate as well as again lower pressure and temperature differences affect the required area, too. As the number of tubes is set according to the maximum velocity in the tubes, more tubes are required in the evaporator for MDM due to higher mass flow rates. As the system pressure in case of combined heat and power generation is lower than in the case of pure electricity generation, a higher part of the area is omitted to the evaporator due to increased heat of vaporisation.

The process parameters in case of the mixture again correspond more to those of MM. The heat duty in the evaporator exceeds that of MM by 22.9%. Heat transfer coefficients are 10.1% higher. An increase in heat transfer area of 38.15% is required. Again, compared to MM, slightly lower temperature differences and pressures are present. The total heat transfer area of preheater (*PH*) and evaporator exceeds that of MM by 14%.

Summing up it has to be stated that the pure heat transfer characteristics of fluids do not necessarily determine the size of the heat exchanger area. Process parameters play an even comparable role. For the investigated cases, the most efficient mixtures require an increase in heat transfer area for the heat input compared to the most efficient pure component of 0.9% and 14%, respectively. This corresponds to an increase in second law efficiency of 0.8% and 2.9%, respectively. As stated in the introduction, the potential of efficiency increase in high temperature ORC is low compared to low temperature applications. Heberle [35] reports an efficiency increase of almost 16% for a geothermal application using mixtures of R245fa and R365mfc. But the required heat transfer area (evaporator + preheater) for the most efficient mixture composition exceeds that of the most efficient pure fluid by 83% [35]. This effect is strongly influenced by the highly reduced heat transfer coefficients of R245fa/R365mfc-mixture compared to their pure components. Reductions compared to the pure fluids of up to nearly 50% are observed [36], whereas for the here investigated siloxane mixtures maximum deviations from the pure component with the lowest heat transfer coefficient are −21.5%. Therefore, it can be concluded, that even though comparably low efficiency increases are possible in high temperature applications, the required additional heat transfer area for mixtures is reduced compared to low temperature systems.

As the present investigations use an additional thermal oil loop, liquid heat transfer coefficients of heat source and ORC-fluid are quite similar. Therefore, the heat transfer coefficient of the ORC-fluid during evaporation can have a strong effect on the required heat transfer area. Regarding applications with direct use of exhaust gas, this effect is diminished or even negligible due to the generally low heat transfer characteristics of the exhaust gas. In these cases, already slight increases in efficiency by using mixtures can justify their application.

Next to the evaporation process, it is necessary to investigate the heat transfer areas of internal heat exchanger and condenser as well. As during internal heat exchange no phase change occurs, degradation of heat transfer coefficients of mixtures are not expected and, therefore, there should be no additional increase in heat transfer area. In contrary, during condensation, reduced heat transfer coefficients in case of mixtures are obvious. Thus, the heat transfer area of the condenser has to be calculated, too. In case of directly air cooled processes, the effect of heat transfer degradation of mixtures is again negligible due to the worse heat transfer characteristics of air. But, regarding the case of combined heat and power generation, where water is used as a cooling medium, the effect of the lower heat transfer coefficients of mixtures can again result in larger heat exchange areas.

5. CONCLUSIONS

The performance of the two linear siloxanes MM and MDM as well as their mixtures in a high temperature ORC are analysed using the example of waste heat from biogas plant. In case of pure electricity generation only slight improvements (1.3%) in second law efficiency can be achieved by using a mixture of 97 wt % MM. An efficiency increase of nearly 3% is obtained for the combined heat and power generation with the mixture 60/40-wt % MM/MDM. As the potential of efficiency increase drops with rising heat source temperature, these results seem reasonable compared to those given in literature.

Next to the thermodynamic analysis, measurements of heat transfer coefficients of the pure siloxanes as well as their mixtures (steps of 20 wt % MM) have been performed. Results show degradations of heat transfer coefficients compared to the ideal values of up to 46%. Kandlikar's correlation is used to describe the experimental data. In order to achieve a good agreement with the real values, the fluid-surface parameter was used to adjust the correlation. Moreover, the severe-diffusion induced region was chosen. By using this adapted correlation, mean deviations of less than 10% between the experimental data and the correlation are present.

Based on all these results, required heat transfer areas of evaporator and preheater are estimated assuming shell-and-tube heat exchangers. Calculations are performed for the case of pure electricity generation as well as combined heat and power generation for the pure components and one mixture. Results show an increased heat transfer area in case of the mixtures of around 1% (pure electricity, 95 wt % MM, efficiency increase: ≈ 1%) and 14% (combined heat and power, 60 wt % MM, efficiency increase: ≈ 3%). Moreover, it is obvious that next to the heat transfer coefficients, process parameters do have a strong influence on the required heat exchange area. Therefore, it cannot be stated that lower heat transfer coefficients lead to higher heat transfer areas than comparing a certain application. It is always necessary to calculate the respective areas under consideration of the (sometimes even just slightly) different operating conditions. However, it can be concluded that siloxanes, as working fluids for high temperature applications, exhibit reduced heat transfer degradation compared to low temperature fluids like for example R245fa/R365mfc. As a result, heat transfer areas are not as additionally increased as in case of geothermal applications and, therefore, the use of mixtures can be beneficial, even though efficiency increases are comparably small.

In order to definitely decide whether the use of mixtures is economically feasible in a certain application, it is certainly necessary to analyse the heat transfer areas of internal heat exchanger and condenser as well. In addition, the performance of turbine and pump has to be investigated, too. Finally, specific problems that can occur by using mixtures, as for example compositions shifts during the process, have to be analysed in detail. Therefore, in future work, analogous investigations (measurements of heat transfer coefficients, identifying of appropriate correlations as well as estimation of required heat transfer areas) will be performed with respect to the condenser. Then, a holistic evaluation of using mixtures in ORC-applications is possible.

ACKNOWLEDGMENTS

The work was partially funded by the Deutsche Forschungsgemeinschaft (DFG) with project No. BR 1713/12. The authors gratefully acknowledge this support.

AUTHOR CONTRIBUTIONS

All authors contributed to this work by collaboration. Theresa Weith is the main author of this manuscript. Florian Heberle assisted in the conceptual design of the study as well as in the experiments. Markus Preißinger particularly contributed to the heat exchanger model. The whole project was supervised by Dieter Brüggemann. All authors revised and approved the publication.

CONFLICTS OF INTEREST

The authors declare no conflict of interest.

REFERENCES

1. Heberle, F.; Preißinger, M.; Brüggemann, D. Zeotropic mixtures as working fluids in Organic Rankine Cycles for low-enthalpy geothermal resources. *Renew. Energy* **2012**, *37*, 364–370.
2. Aghahosseini, S.; Dincer, I. Comparative performance analysis of low-temperature Organic Rankine Cycle (ORC) using pure and zeotropic working fluids. *Appl. Therm. Eng.* **2013**, *54*, 35–42.
3. Liu, Q.; Duan, Y.; Yang, Z. Effect of condensation temperature glide on the performance of organic Rankine cycles with zeotropic mixture working fluids. *Appl. Energy* **2014**, *115*, 394–404.
4. Preißinger, M.; Heberle, F.; Brüggemann, D. Advanced Organic Rankine Cycle for geothermal application. *Int. J. Low Carbon Technol.* **2013**, *8*, i62–i68.
5. Chen, H.; Goswami, D.; Rahman, M.; Stefanakos, E. A supercritical Rankine cycle using zeotropic mixture working fluids for the conversion of low-grade heat into power. *Energy* **2011**, *36*, 549–555.
6. Baik, Y.-J.; Kim, M.; Chang, K.-C.; Lee, Y.-S.; Yoon, H.-K. Power enhancement potential of a mixture transcritical cycle for a low-temperature geothermal power generation. *Energy* **2012**, *47*, 70–76.
7. Yang, K.; Zhang, H.; Wang, Z.; Zhang, J.; Yang, F.; Wang, E.; Yao, B. Study of zeotropic mixtures of ORC (organic Rankine cycle) under engine various operating conditions. *Energy* **2013**, *58*, 494–510.
8. Zhang, J.; Zhang, H.; Yang, K.; Yang, F.; Wang, Z.; Zhao, G.; Liu, H.; Wang, E.; Yao, B. Performance analysis of regenerative organic Rankine cycle (RORC) using the pure working fluid and the zeotropic mixture over the whole operating range of a diesel engine. *Energy Convers. Manag.* **2014**, *84*, 282–294.

9. Angelino, G.; Colonna, P. Multicomponent Working Fluids for Organic Rankine Cycles (ORCs). *Energy* **1998**, *23*, 449–463.

10. Angelino, G.; Colonna, P. Air cooled siloxane bottoming cycle for molten carbonate fuel cells. In Proceedings of the Fuel Cell Seminar, Portland, OR, USA, 30 October 2000; pp. 667–670.

11. Dong, B.; Xu, G.; Cai, Y.; Li, H. Analysis of zeotropic mixtures used in high-temperature Organic Rankine cycle.*Energy Convers. Manag.* **2014**, *84*, 253–260.

12. Chys, M.; van den Broek, M.; Vanslambrouck, B.; de Paepe, M. Potential of zeotropic mixtures as working fluids in organic Rankine cycles. *Energy* **2012**, *44*, 623–632.

13. Kedzierski, M.; Kim, J.; Didion, D. Causes of the apparent heat transfer degradation for refrigerant mixtures. In Proceedings of the ASME/AIChe/ANS National Heat Transfer Conference, San Diego, CA, USA, 9–12 August 1992; pp. 149–158.

14. Stephan, K. Heat transfer in boiling of mixtures. In Proceedings of the Seventh International Heat Transfer Conference, Munich, Germany, 6–10 September 1982; pp. 59–81.

15. Bivens, D.; Yokozeki, A. Heat transfer of refrigerant mixtures. In Proceedings of the International Refrigeration and Air Conditioning Conference, Purdue University, West Lafayette, IN, USA, 14–17 July 1992; pp. 141–148.

16. Jung, D.; Song, K.; Ahn, K.; Kim, J. Nucleate boiling heat transfer coefficients of mixtures containing HFC32, HFC125, and HFC134a. *Int. J. Refrig.* **2003**, *26*, 764–771.

17. Li, M.; Dang, C.; Hihara, E. Flow boiling heat transfer of HFO1234yf and R32 refrigerant mixtures in a smooth horizontal tube: Part I. Experimental investigation. *Int. J. Heat Mass Transf.* **2012**, *55*, 3437–3446.

18. Choi, T.; Kim, Y.; Kim, M.; Ro, S. Evaporation heat transfer of R-32, R-134a, R-32/134a, and R-32/125/134a inside a horizontal smooth tube. *Int. J. Heat Mass Transf.* **2000**, *43*, 3651–3660.

19. Tibiriçá, C.; Ribatski, G. Flow boiling heat transfer of R134a and R245fa in a 2.3 mm tube. *Int. J. Heat Mass Transf.* **2010**,*53*, 2459–2468.

20. Ong, C.; Thome, J. Flow boiling heat transfer of R134a, R236fa and R245fa in a horizontal 1.030 mm circular channel.*Exp. Therm. Fluid Sci.* **2009**, *33*, 651–663.

21. Fernández, F.; Prieto, M.; Suárez, I. Thermodynamic analysis of high-temperature regenerative organic Rankine cycles using siloxanes as working fluids. *Energy* **2011**, *36*, 5239–5249.

22. Lai, N.; Wendland, M.; Fischer, J. Working fluids for high-temperature organic Rankine cycles. *Energy* **2011**, *36*, 199–211.

23. Heberle, F.; Preißinger, M.; Weith, T.; Brüggemann, D. Experimental Investigation of Heat Transfer Characteristics and Thermal Stability of Siloxanes. Available online: http://www.asme-orc2013.nl/uploads/File/PPT%20082.pdf(accessed on 21 June 2014).

24. The Dow Chemical Company. Dowtherm G, Product Information. Available online: http://msdssearch.dow.com/PublishedLiteratureDOWCOM/dh_0033/0901b803800337f0.pdf?filepath=heattrans/pdfs/noreg/176-01466.pdf&fromPage=GetDoc (accessed on 19 May 2014).

25. Aspen Technology, Incorporation. *Aspen One V 7, Process Optimization for Engineering, Manufacturing, and Supply Chain, Aspen Plus V 7.3*; Aspen Technology: Burlington, ON, Canada, 2011.

26. Aspen Technology, Incorporation. *Aspen Physical Property System. Physical Property Methods. Documentation to Aspen Plus V 7.3*; Aspen Technology: Burlington, ON, Canada, 2011.

27. Lemmon, E.W.; Huber, M.L.; McLinden, M.O. *NIST Standard Reference Database 23: Reference Fluid Thermodynamic and Transport Properties-REFPROP, Version 9.1*; National Institute of Standards and Technology, Standard Reference Data Program: Gaithersburg, MD, USA, 2013.

28. Weith, T.; Heberle, F.; Brüggemann, D. Experimental Investigation of Flow Boiling Characteristics of Siloxanes and Siloxane Mixtures in a Horizontal Tube. In Proceedings of the International Symposium on Convective Heat and Mass Transfer, CONV-14, Kusadasi, Turkey, 8–13 June 2014.

29. Kandlikar, S. Boiling heat transfer with binary mixtures: Part II—Flow boiling in plain tubes. *J. Heat Transf.* **1998**, *120*, 388–394.

30. Podhorsky, M.; Krips, H. *Wärmetauscher—Aktuelle Probleme der Konstruktion und Berechnung*, 2nd ed.; Vulkan-Verlag: Essen, Germany, 1999. (In German)

31. Wolf, P.; Kirchner, G.O. Konstruktive Hinweise für den Bau von Wärmeübertragern. In *VDI-Wärmeatlas*, 10th ed.; Verein Deutscher Ingenieure: Berlin, Germany, 2013. (In German)

32. Roetzel, W.; Spang, B.C. Berechnung von Wärmeübertragern. In *VDI-Wärmeatlas*, 10th ed.; Verein Deutscher Ingenieure: Berlin, Germany, 2013. (In German)

33. Gnielinski, V. Ein neues Berechnungsverfahren für die Wärmeübertragung im Übergangsbereich zwischen laminarer und turbulenter Rohrströmung. *Forsch. Geb. Ing.* **1995**, *61*, 240–248. (In German).

34. MathWorks, Inc. *MATLAB R2010b*; MathWorks, Inc.: Natick, MA, USA, 2010.

35. Heberle, F. Untersuchungen zum Einsatz von zeotropen Fluidgemischen im Organic Rankine Cycle für die geothermische Stromerzeugung. Ph.D. Thesis, Universität Bayreuth, Bayreuth, Germany, 2013. (In German).

36. Heberle, F.; Brüggemann, D. Pool boiling heat transfer coefficients of R245fa, R365mfc and their mixtures. In Proceeding of the 8th World Conference on Experimental Heat Transfer, Fluid Mechanics, and Thermodynamics, Lisbon, Portugal, 16–20 June 2013.

4
CHAPTER

Theoretical Research on R245fa Condensation Heat Transfer inside a Horizontal Tube

Shengchun Liu*, Yan Huo, Zhaowei Liu, Lan Li, Jinghong Ning

Tianjin Key Laboratory of Refrigeration technology, Tianjin University of Commerce, Tainjin, China

ABSTRACT

Numerical simulation on R245fa condensation inside an inner diameter of 8 mm horizontal tube is researched in this paper. The effect of variation in velocity, condensation temperature and superheat of inlet steam and variation in cooling water temperature on heat transfer coefficient are investigated as a parametric study. Condensation process of steam has been successfully modeled by applying a user defined function (UDF) added to the commercial computational fluid dynamics (CFD) package. By analyzing the corresponding condensate contours and the curves of local heat transfer coefficient, the relationships between condensation heat transfer coefficient and various parameters of R245fa inside horizontal tube are obtained. It shows that the heat transfer coefficient increases by the increase in velocity, condensation temperature and superheat of inlet steam and the decrease in cooling water temperature. The errors between the heat transfer coefficient of simulation result and model of Wang and Shah are within ±30%. The parametric study will provide the basis for designing efficient heat exchangers of R245fa.

Keywords: Condensation Heat Transfer, R245fa, Numerical Simulation, Horizontal Tube

1. INTRODUCTION

Under the dual pressures of energy and environmental issues, developing environmentally friendly refrigerant becomes one of the three major issues facing the refrigeration and air conditioning industry. In recent years, R245fa is an environmentally friendly substituted refrigerant, which is researched by international experts. It has characteristics of excellent thermal resistance, low toxicity, no chlorine, and no damaging effects on the ozone layer. And comparing with R134a, it has a lower pressure level [1] . It is more stable used in some specific situations. As a working fluid of high critical temperature among the zero ODP refrigerants, R245fa represents a substitute for R123 and R11 used in high temperature heat pumps, in jet-pump refrigerators, in centrifugal chillers and in heat recovery power cycles [2] . And R245fa shows high power-generating ability and high economical efficiency in the low-temperature range [3] . R245fa as working fluid in the low-temperature solar power rankine cycle system is feasible and the performance is acceptable [4] . Eames's experiment [5] results demonstrate that low-grade heat can be used to drive efficient jet-pump refrigeration systems, using R245fa as their working fluid, for high temperature applications, such as air conditioning.

It is very important to research the heat transfer characteristics of refrigerants R245fa. Zou et al. [6] used R245fa, R134a, R410A, and R32 to explore the effects of fluid properties on refrigerant distribution in the vertical header of a reversible micro-channel heat exchanger. By comparing the results of experiments, R245fa was found to the best distribution. It is because of the low steam density and high liquid density of R245fa. Many scholars have studied the flow boiling heat transfer of R245fa. In Vakili-Farahani et al.'s experiment [7] with refrigerants R1234ze and R245fa, the heat transfer coefficient increases with rising of saturated temperature, heat flux and mass flux. Tibiriçá et al. [8] did experiment in a horizontal 2.3 mm I.D. stainless steel tube utilizing working fluids R134a and R245fa. It was concluded that the heat transfer coefficient was related to heat flux, mass velocity and quality. Ong et al. [9] , Pike-Wilson et al. [10] and Agostini et al. [11] researched on the flow boiling patterns, pressure drop and heat transfer characteristic of R245fa in a tube. Researching on condensation heat transfer inside tubes is another important aspect of heat transfer characteristics, and the literature on condensation of R245fa is few.

In this paper, condensation heat transfer is simulated by using the commercial software CFD, VOF model and UDF (User-Defined Functions). Numerical simulation on R245fa steam condensing in the inner diameter of 8 mm horizontal tube and analysis is conducted in terms of steam flow rate, cooling water temperature, condensation temperature and superheat. By analyzing the corresponding condensate contours and the curves of local heat transfer coefficient, the relationships between condensation heat transfer coefficient and various parameters inside horizontal tube are obtained. It can provide the basis for designing efficient heat exchangers of R245fa.

2. NUMERICAL METHODOLOGY

A horizontal tube with inner diameter of 8 mm, length 500 mm is considered. And the model coordinates is shown in Figure 1. The origin of the coordinate system is at the center of the interface, the pipe longitudinal direction of the X-axis, the vertical direction for the Z-axis, the influence of gravity is considered in the simulation, and the acceleration of gravity is 9.806 N/kg, the Z-axis is parallel to direction of gravity.

According to pipe geometry, the model is meshed by size of 1 mm using ring hexahedral structured grid. The refrigerant flows into the pipe, and condensation occurs when it comes across the colder wall surface.

The condensation film attached to the wall, and the vapor is in the central area of pipe. Considering the impact of the boundary layer near the wall, Boundary layer near the wall is meshed, and the initial boundary layer mesh size is 0.2 mm, growth ratio of 1.2, the boundary layer mesh layers of 6. The meshing results are shown in Figure 2 and Figure 3.

Figure 4 is a simplified schematic diagram of the simulated section of the tube and boundary conditions, Inlet and outlet boundary condition are set to velocity-inlet and outflow respectively. And constant temperature of the pipe wall is also set.

Many researchers apply VOF model to go on investigation of two phase flow. Such as literature [12] -[14] using it to study boiling and literature [15] [16] using it to research condensation. In this paper, using commercial software Fluent 14.0, multiphase flow model chooses VOF model, calculating as follows within controlling cell.

Within each cell, the volume fraction of phases is 1 for gas-liquid two-phase flow, and as the presence of the following formula:

$$\alpha_l + \alpha_v = 1 \tag{1}$$

Density ρ within each cell, the thermal conductivity λ, the dynamic viscosity μ and thermodynamic energy E are calculated as follows:

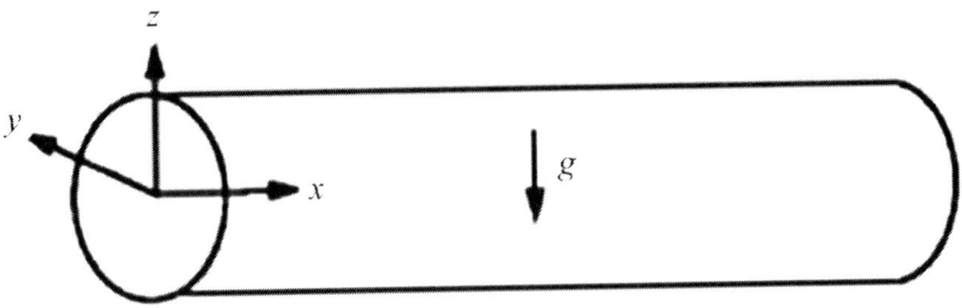

Figure 1. The model coordinates.

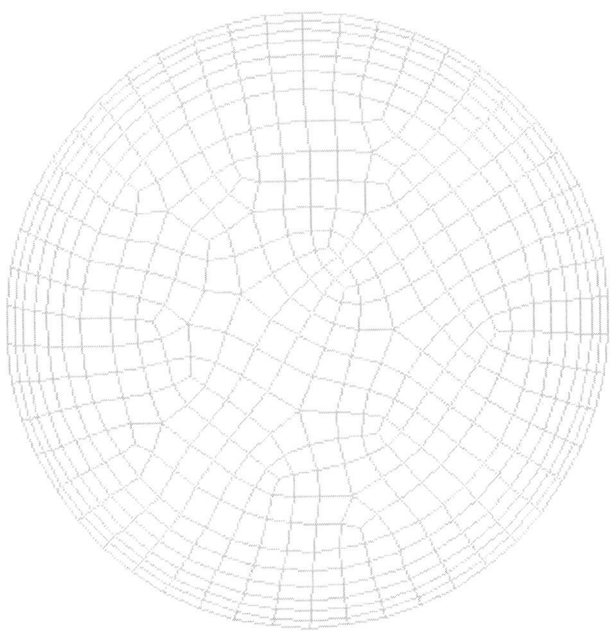

Figure 2. The sectional grid.

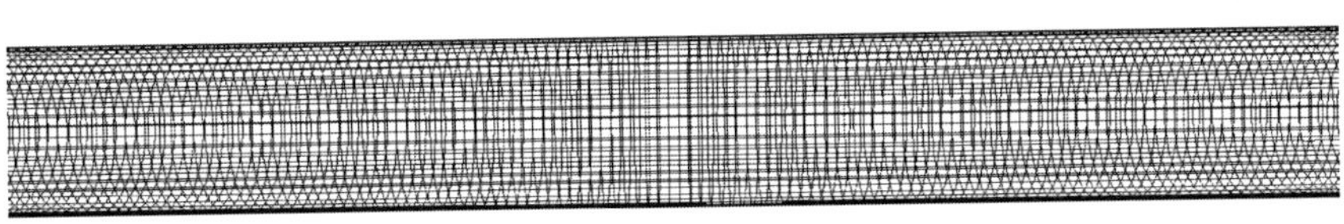

Figure 3. The grid of wall surface.

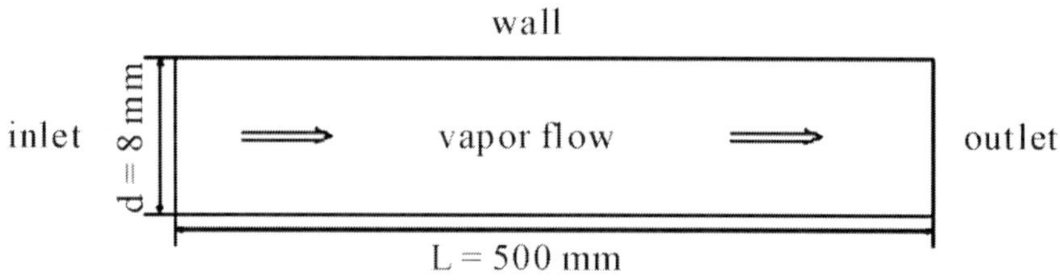

Figure 4. Simplified schematic diagram and boundary conditions.

$$\rho = \alpha_l \rho_l + (1 - \alpha_l)\rho_v \tag{2}$$

$$\lambda = \alpha_l \lambda_l + (1 - \alpha_l)\lambda_v \tag{3}$$

$$\mu = \alpha_l \mu_l + (1 - \alpha_l)\mu_v \tag{4}$$

$$E = \frac{\alpha_l \rho_l E_l + (1 - \alpha_l)\rho_V E_v}{\alpha_l \rho_l + (1 - \alpha_l)\rho_V} \tag{5}$$

And continuity equation, momentum equation and energy equation can be expressed as follows:

$$\frac{\partial \alpha_l}{\partial \tau} + u \cdot \nabla \alpha_l = \frac{S_l}{\rho_l} \tag{6}$$

$$\frac{\partial \alpha_v}{\partial \tau} + u \cdot \nabla \alpha_v = \frac{S_v}{\rho_v} \tag{7}$$

$$\frac{\partial(\rho u)}{\partial \tau} + \nabla \cdot (\rho u u) = -\nabla p + \nabla \cdot \left[\mu\left(\nabla u + \nabla u^T\right)\right] + \rho g + F \tag{8}$$

$$\frac{\partial(\rho E)}{\partial\tau}+\nabla\cdot\left[u(\rho E+p)\right]=\nabla\cdot\left(\lambda_{eff}\nabla T\right)+Q \tag{9}$$

In the present CFD model, the liquid evaporation and vapor condensation effects can be considered by incorporating mass and energy sources into the continuity and energy equations, respectively. The source term for the energy equation can be obtained by multiplying the rate of mass transfer by the latent heat. In a CFD simulation, the source terms act on the whole fluid domain, so the phase change effect both at liquid surface and in the bulk region can be considered simultaneously. All of the source terms are driven by the cell temperature, T, and implemented via customized UDF. Based on the following temperature regimes, the mass transfer can be described as follows [17] :

If $T \geq T_{sat}$ (evaporation),

$$m=\varphi\alpha_l\rho_l\frac{\left|T-T_{sat}\right|}{T_{sat}} \tag{10}$$

If $T < T_{sat}$ (condensation),

$$m=\varphi\alpha_v\rho_v\frac{\left|T-T_{sat}\right|}{T_{sat}} \tag{11}$$

Using the k-ε turbulence model, the formats of simulation variables to select are shown in Table 1, and the convergence criterion of control equations are shown in Table 2.

3. RESULTS AND DISCUSSION

Refrigerants R245fa inlet steam is saturated, the condensation temperature is 314 K, the wall temperature is 293 K, and the inlet steam flow rate is 6 m/s, 10 m/s, 14 m/s and 18 m/s respectively. The changes of film thickness of the inner tube and heat transfer coefficient are shown in Figure 5.

From above curves, it is concluded that the local heat transfer coefficient increased, the thickness of condensate film decreased and the rate of increase of condensation film thickness slow down, as the inlet steam flow

Table 1. Format of variables.

Variables	Format
Solver	Pressure based
Time	Steady
VOF format	Implicit
Pressure-velocity coupling	SIMPLE
Pressure	Body force weighted
Momentum	Second-order upwind
Volume fraction	First-order upwind
Turbulent kinetic energy	Second-order upwind
Turbulent dissipation rate	Second-order upwind
Energy	Second-order upwind

Table 2. The convergence criterion of control equations.

Equations	Residuals
Continuity	1×10^{-3}
Momentum	1×10^{-3}
Energy	1×10^{-3}
Turbulence	1×10^{-3}

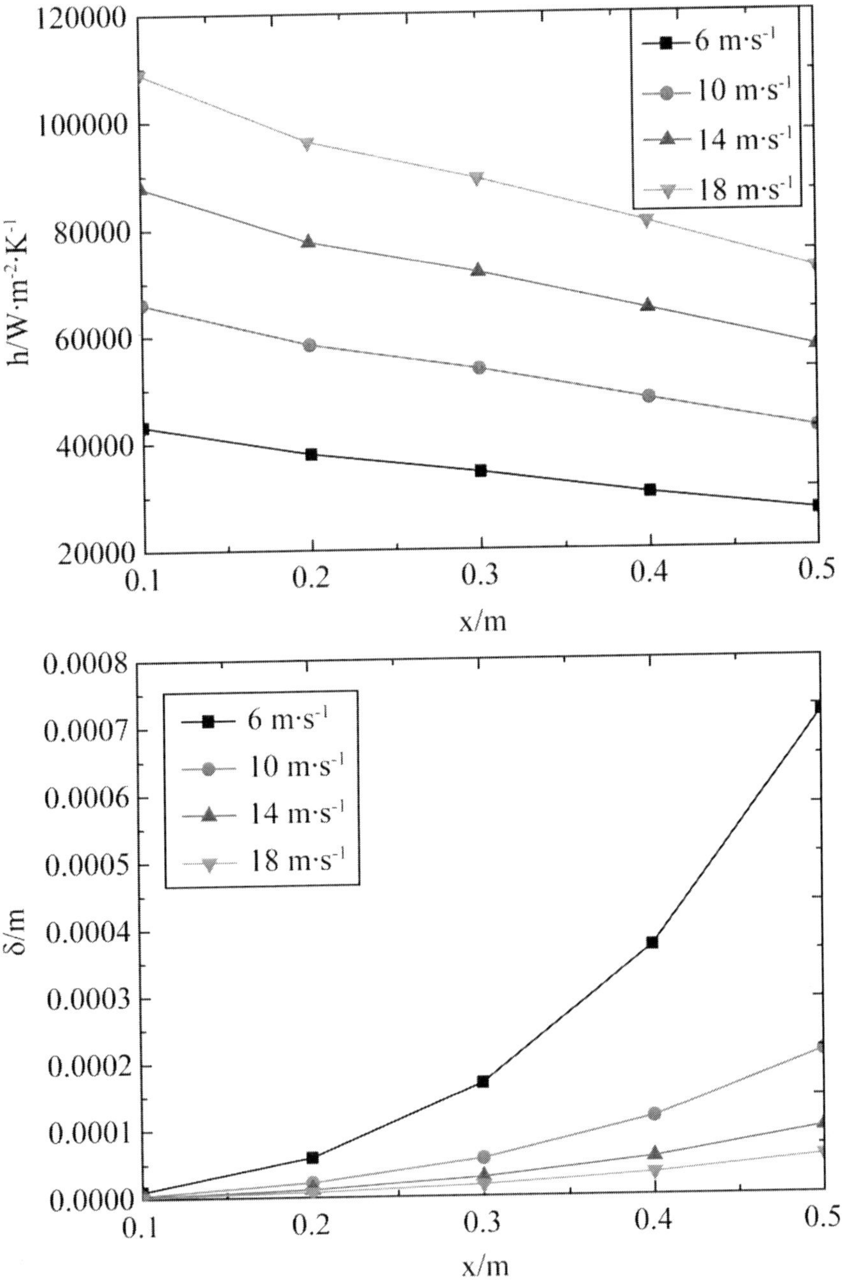

Figure 5. Curves under different flow rates.

rate increases. And condensation heat transfer coefficient changes dramatically, when refrigerant steam flows into tube. At the Steam flow rate of 6 m/s, condensate begins to increase rapidly in 0.3 m at growth gradually accelerated. As the flow rate increases, the changing rate of local heat transfer coefficient increases. The faster steam flow rate, the stronger the shear is, the more slowly the condensation film thickness increases.

The inlet velocity is 10 m/s, and the contours of liquid volume fraction are shown in Figure 6 andFigure 7, respectively. The surface heat transfer coefficient is shown in Figure 8. The surface heat transfer coefficient increases at first and then decreases, therefore there is a maximum, and the rate of increase becomes faster close to the entrance of steam. It can be interpret as part of the steam condensed to form droplets at the inlet; with the process of condensation, droplets become larger, increasing the heat transfer area. Then the droplets become liquid film, film formation increases heat resistance, making the heat transfer coefficient decreases.

As shown in Figure 7 and Figure 8, because the steam temperature is higher than that of pipe wall, the steam condenses on the wall. With the working fluid flowing and the occurrence of condensation, the film thickness increased along direction of flowing. Due to gravity, most of liquid accumulates in the bottom of the pipe. The condensation heat transfer coefficient changed dramatically as refrigerant steam flowed into tube. Then, the heat transfer coefficient reduced, the heat transfer coefficient is further reduced quickly at the position of about 0.3 m.

Refrigerants R245fa inlet steam is saturated, the inlet steam flow rate is 10 m/s, the condensation temperature is 314 K, and the wall temperature was 288 K, 291 K, 293 K and 295 K respectively. The changes of film thickness of the inner tube and heat transfer coefficient are shown in Figure 9. The lower the temperature of the tube wall, the great heat transfer coefficient is, the more condensed liquid is. Analysis of the causes is that the lower tube wall temperature, the greater heat temperature difference, and the larger condensation heat transfer. Condensate begins to increase rapidly at 0.3 m and the thickening rate of film increased, heat transfer coefficient decreases rapidly.

Refrigerants R245fa inlet steam is saturated, the inlet steam flow rate is 10 m/s, the wall temperature is 293 K, and the condensation temperature is 308 K, 313 K and 318 K, respectively. The changes of film thickness of the inner tube and heat transfer coefficient are shown in Figure 10.

Changes of the film thickness and heat transfer coefficient along tube are similar to the cooling water temperature trends under condensation temperatures. The larger the heat transfer coefficient, the thicker the condensation film. The rate of condensation film changing become fast with the rising of condensation temperature. These can be expressed as the higher condensation temperature, the greater the temperature difference, there is a growing rate of condensation. That condensate begins to increase rapidly at 0.3 m and the rate of increase gradually accelerated can also be obtained.

Refrigerants R245fa inlet steam is superheated with 1 K, 5 K, 10 K and 15 K. The inlet steam flow rate is 10 m/s, the wall temperature is 293 K, and the condensation temperature is 314 K, respectively. The changes of film thickness and heat transfer coefficient are shown in Figure 11.

When the superheat increases, the heat transfer coefficient becomes larger and the amount of condensate reduces. Superheat temperature difference leads to heat transfer coefficient increasing, but the amount needed for

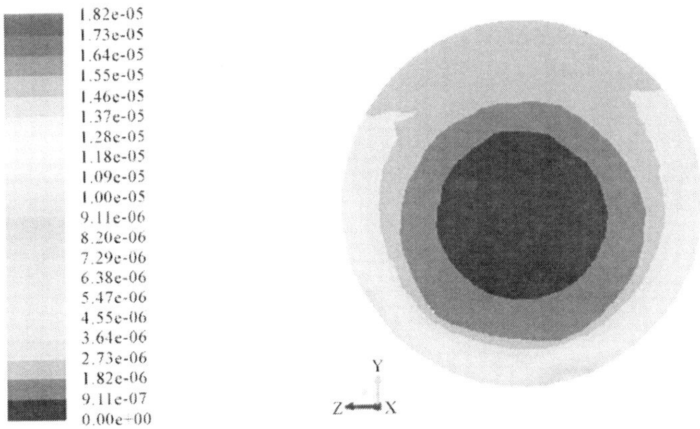

x = 0.2 m

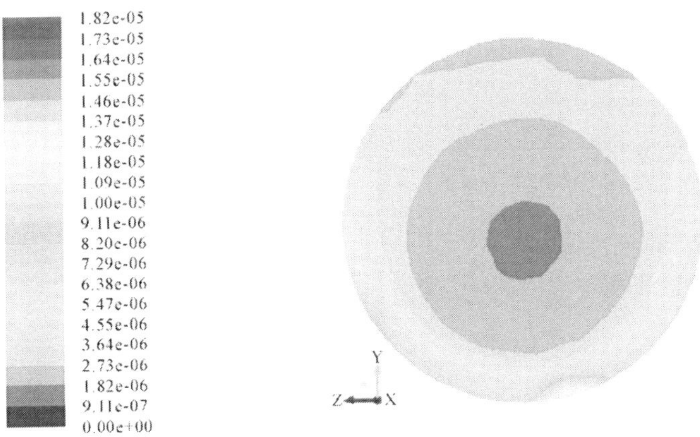

x = 0.3 m

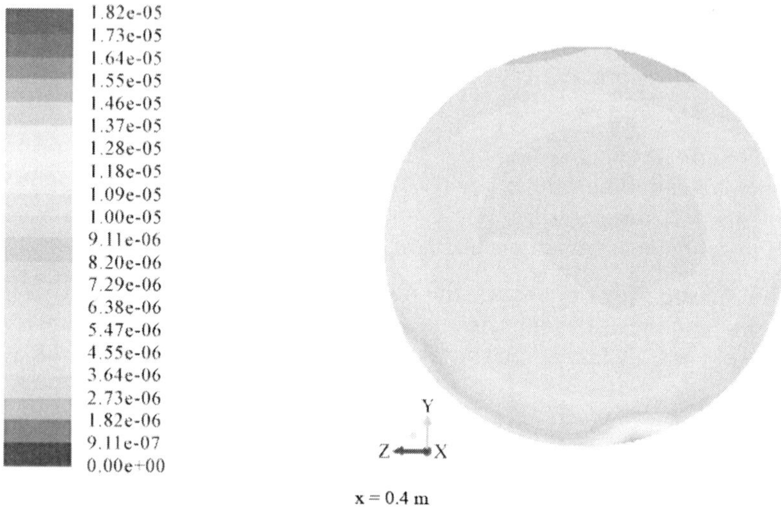

Figure 6. The contours of liquid volume fraction at different positions.

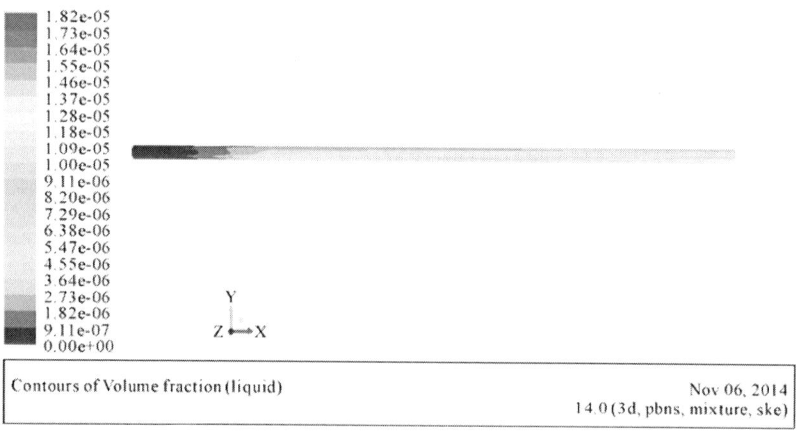

Figure 7. The contours of liquid volume fraction along the tube at the velocity of 10 m/s.

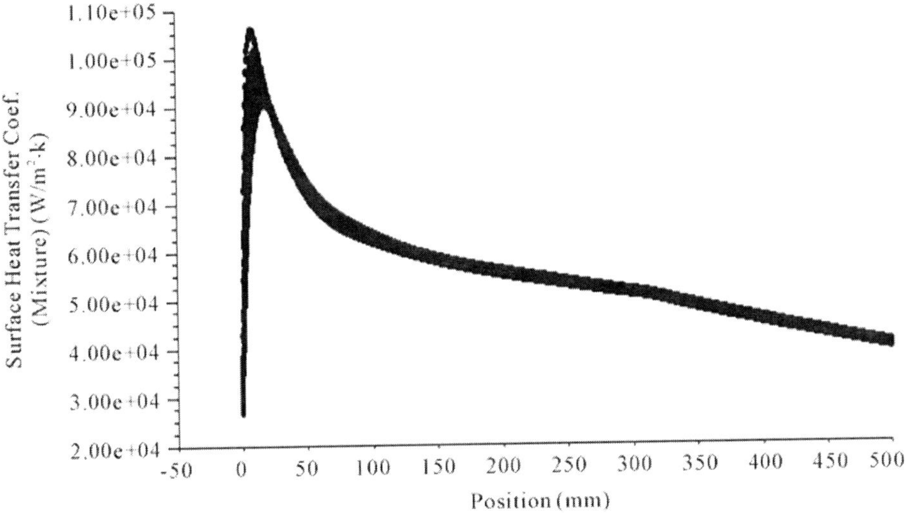

Figure 8. The surface heat transfer coefficient along the tube at the velocity of 10 m/s.

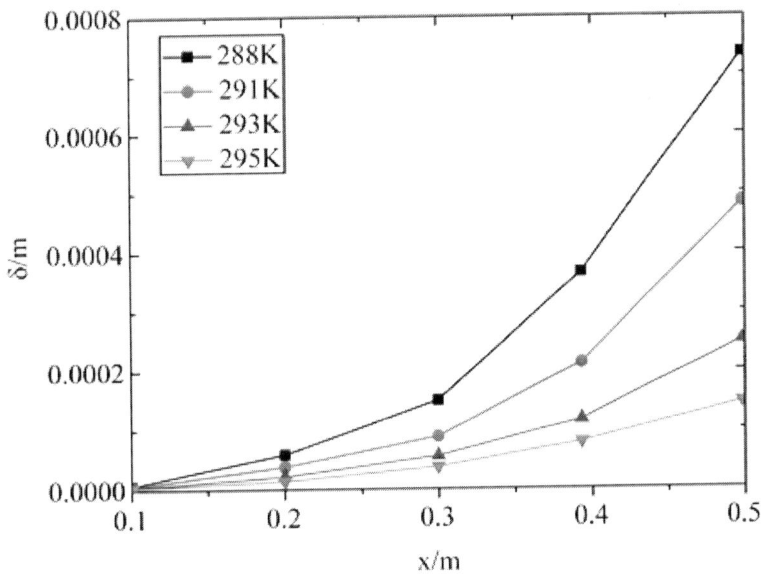

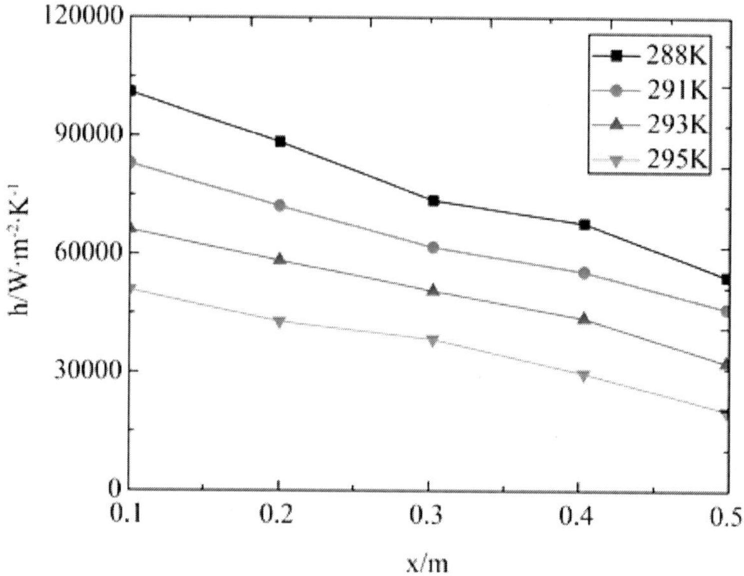

Figure 9. Curves under different wall temperature.

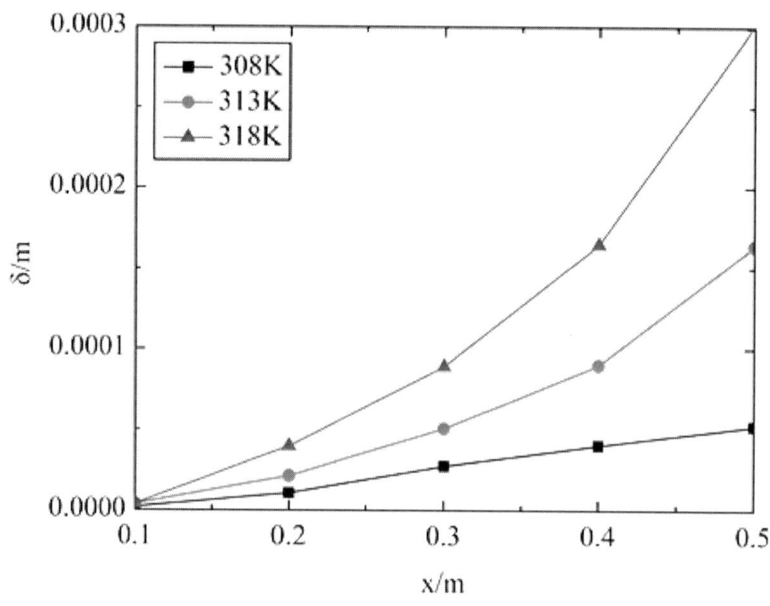

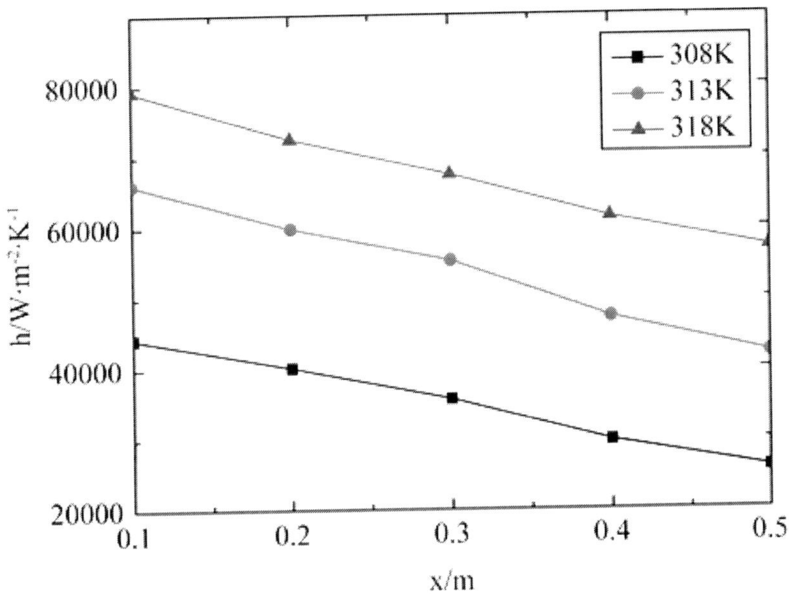

Figure 10. Curves under different condensation temperature.

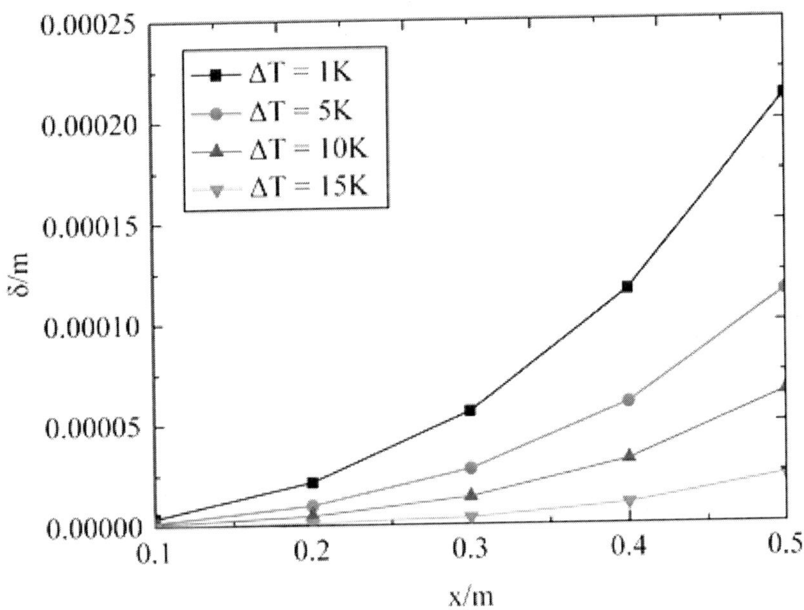

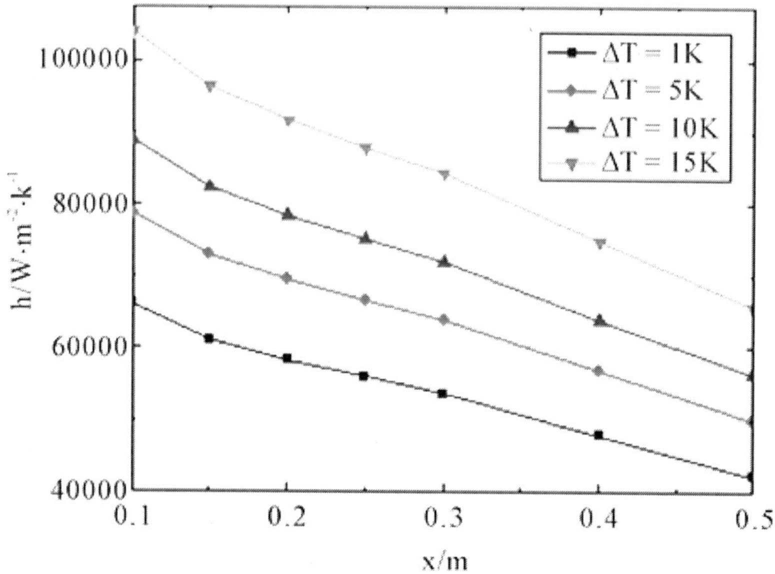

Figure 11. Curves under different superheat.

the cold steam condensation increases. Film thickness starts to increase rapidly at 0.3 m, the growth rate increases gradually, and the curve of corresponding heat transfer coefficient has the tendency to decline rapidly.

As the inlet steam flow rateis 10 m/s, the wall temperature is 293 K, and condensation temperature is 308 K, the errors between the heat transfer coefficient of simulation result and model of Wang [18] and Shah [19] are within ±30%.

4. CONCLUSIONS

By simulation on the process of R245fa condensing inside a horizontal tube and analysis of simulation results, various parameters have great influences on the condensation heat transfer coefficient. The simulation results show that when the inner diameter of tube is 8 mm, the heat transfer coefficient and film thickness have significant changes after the position of 0.3 m. Considering the changes of film thickness and heat transfer coefficient under different conditions, some relevant measures can be taken to reduce the film thickness at about 0.4 m and to enhance heat transfer in this case. The changing regulars of film thickness and heat transfer coefficient in different diameter can be got by using the similar method of simulation in the paper. Then, the position can be obtained to enhance heat transfer. And it can provide the basis for designing efficient heat exchangers of R245fa.

By comparing the different factors on the condensation heat transfer, it can be seen that the ways of influences of cooling water temperature, condensing temperature and steam flow rate on the condensation heat transfer are different. The increasing of team flow rate results in enhanced shear, and condensation film is not stable in the presence of the tube, so film thickness reduces, and heat transfer coefficient increases. The wall temperature decreases or condensing temperature rises, and the temperature difference increases, so that the heat transfer coefficient increases. The heat transfer coefficient of simulation results is close to the model of Wang and Shah. Thus, the model and method of simulation are suitable for process of condensation.

ACKNOWLEDGEMENTS

Shengchun Liu, Yan Huo, Zhaowei Liu, Lan Li, Jinghong Ning (863 Program) (Grant No. 2012AA053001).

REFERENCES

1. Wang, X.D., Zhao, L. and Wang, J.L. (2000) Experimental Investigation on the Low-Temperature Solar Rankine Cycle System Using R245fa. Energy Conversion and Management, 52, 946-952.
2. Angelino, G. and Invernizzi, C. (2003) Experimental Investigation on the Thermal Stability of Some New Zero ODP Refrigerants. International Journal of Refrigeration, 26, 51-58
3. Saleh, B., Koglbauer, G., Wendland, M. and Fischer, J. (2007) Working Fluids for Low-Temperature Organic Rankine Cycles. Energy, 32, 1210-1221.
4. Wang, X.D., Zhao, L., Wang, J.L., Zhang, W.Z., Zhao, X.Z. and Wu, W. (2010) Performance Evaluation of a Low- Temperature Solar Rankine Cycle System Utilizing R245fa. Solar Energy, 84, 353-364.
5. Eames, I.W., Ablwaifa, A.E. and Petrenko, V. (2007) Results of an Experimental Study of an Advanced Jet-Pump Refrigerator Operating with R245fa. Applied Thermal Engineering, 27, 2833-2840.

6. Zou, Y. and Hrnjak, P.S. (2014) Effects of Fluid Properties on Two-Phase Flow and Refrigerant Distribution in the Vertical Header of a Reversible Microchannel Heat Exchanger—Comparing R245fa, R134a, R410A, and R32. Applied Thermal Engineering, 70, 966-976.

7. Vakili-Farahani, F., Agostini, B. and Thome, J.R. (2013) Experimental Study on Flow Boiling Heat Transfer of Multiport Tubes with R245fa and R1234ze(E). International Journal of Refrigeration, 36, 335-352.

8. Tibiriçá, C.B. and Ribatski, G. (2010) Flow Boiling Heat Transfer of R134a and R245fa in a 2.3 mm Tube. International Journal of Heat and Mass Transfer, 53, 2459-2468.

9. Ong, C.L. and Thome, J.R. (2009) Flow Boiling Heat Transfer of R134a, R236fa and R245fa in a Horizontal 1.030 mm Circular Channel. Experimental Thermal and Fluid Science, 33, 651-663.

10. Pike-Wilson, E.A. and Karayiannis, T.G. (2014) Flow Boiling of R245fa in 1.1 mm Diameter Stainless Steel, Brass and Copper Tubes. Experimental Thermal and Fluid Science, 59, 166-183.

11. Agostini, B., Thome, J.R., Fabbri, M., Michel, B., Calmi, D. and Kloter, U. (2008) High Heat Flux Flow Boiling in Silicon Multi-Microchannels—Part II: Heat Transfer Characteristics of Refrigerant R245fa. International Journal of Heat and Mass Transfer, 51, 5415-5425.

12. Rabha, S.S. and Buwa, V.V. (2009) Volume-of-Fluid (VOF) Simulations of Rise of Single/Multiple Bubbles in Sheared Liquids. Chemical Engineering Science, 65, 527-537.

13. Yang, Z., Peng, X.F. and Ye, P. (2008) Numerical and Experimental Investigation of Two Phase Flow during Boiling in a Coiled Tube. International Journal of Heat and Mass Transfer, 51, 1003-1016.

14. Huang, M., Yang, Z., Duan, Y.Y. and Lee, D.J. (2011) Bubble Growth for Boiling Bubbly Flow for R141b in a Serpentine Tube. Journal of the Taiwan Institute of Chemical Engineers, 42, 727-734.

15. Chen, S.H., Yang, Z., Duan, Y.Y., Chen, Y. and Wu, D. (2014) Simulation of Condensation Flow in a Rectangular Microchannel. Chemical Engineering and Processing, 76, 60-69.

16. Da Riva, E. and Del Col, D. (2012) Numerical Simulation of Laminar Liquid Film Condensation in a Horizontal Circular Minichannel. Journal of Heat Transfer, 134, Article ID: 051019.

17. Wang, L., Li, Y.Z., Li, C. and Zhao, Z.X. (2013) CFD Investigation of Thermal and Pressurization Performance in LH2 Tank during Discharge. Cryogenics, 57, 63-73.

18. Wang, W.W.W., Radcliff, T.D. and Christensen, R.N. (2002) A Condensation Heat Transfer Correlation for Millimeter-Scale Tubing with Flow Regime Transition. Experimental Thermal and Fluid Science, 26, 473-485.

19. Shah, M.M. (2010) Heat Transfer during Condensation inside Small Channels: Applicability of General Correlation for Macrochannels. Proceedings of the 14th International Heat Transfer Conference, Washington DC, 8-13 August 2010, 298-303.

5
CHAPTER

Calculation Methods for Heating and Ventilation Systemof Electrical Machines

Otilia Nedelcu and Corneliu Ioan Sălişteanu

[1] *Department of Electronics, Telecommunications and Energy Engineering, Valahia University of Targoviste, Romania*

1. INTRODUCTION

To design as accurately electrical machines it is necessary as the cold machine to be efficient and to determine values and distribution of temperature in the machine (along the rotor). For this, must do a complete and accurate analysis of temperature field in steady state (or in transient regime) which requires the knowledge and location of the temperature loss amounts, the knowledge of the velocity distribution of the cooling fluid in the cooling channels different of the machine, the determination of thermal flux values and the distribution of these values in some component parts of the electric machine.

Considering the analogy between a ventilation network of cooling and a nonlinear resistive circuit, the flows of the cooling agent of an electric machine can be determined.

The aim of thermal computation of electric machines is necessary to establish heating, temperature overrunning or overheating in some parts of the machine from the environment temperature at a given operating condition of the machine.

The service life of the insulations used in electric machines is limited heating which greatly decreases in the same time with temperature increasing. Heating determined is stabilized, because the electrical machine which will be discussedhas a long-lasting operating regime. Inside the electric machine a heat quantity is produced which has to be exhausted to outside for limiting the excessive heating, heating due to the losses which occur in the active materials.

2. HEAT TRANSFER

The heat transfer is the science of spontaneous and irreversible processes of heat propagation in space and is the exchange of thermal energy between two objects, two regions of the same part and respectively two fluids, as a result of a difference of temperature between two fluids.

In technical, the heat transfer must be:

- more intense (e.g. steam boilers, the optimal exchange in heat exchangers);
- Stopped (e.g. steam by thermal insulation).

The heat is the transfer of energy between phisical-chemical systems or between different parts of the same systems, if there is a transformation where is not performed any mechanical work.

The heat transfer deals with the processes where the thermical energy at higher parameters is changed in thermal energy to lower parameters, usually, the parameter which appreciates the heat quality is the temperature, defined as a global measure of processes' intensity which determines the internal energy of an object (thermal agitation of molecules in liquid and gases, atoms vibration and free electrons motion in metals, etc.).

The heat exchange follows thermodynamics principles: **the first principleof thermodynamics**expresses the energy conservation rule and it establishes the quantitative relationship between heat and mechanical work, and allows integral transformation of heat in mechanical work and the reverse;**the second principle of thermodynamics** states the necessary conditions for transforming heat in mechanical work and establishes natural meaning of heat propagation, always from the source with higher temperature to the source with lower temperature.

The main objectives of the heat transfer are: first, the determination or ensuring the amount of heat exchanged per time unit in fixed temperature conditions, and second, checking the compatibility of materials used with the regime of temperature faced by, in determining the temperature field. We can say that an exchange of heat device is an optimal solution from thermic, hydraulic, mechanic, economic and reliability points of view, mainly, the heat transfer being the determining factor. We should add to these, finding the enhancing methods and procedures or, in certain cases, heat transfer breaking. [5]

Conduction, radiation and convection are distinct forms of heat transfer. As defined above, only**conduction** and **radiation** are heat exchange processes due solely to the difference in temperature,**convection** is a more complex process, necessarily involving the mass transfer. Because, the convection makes energy transfer from regions with higher temperature in regions with lower temperature, it was generally accepted „heat transfer by convection" as being the third mode of heat exchange.

Basic sizes in heat transfer

The temperature field. The status scalar parameter, the temperature, in any point of space M(x, y, z) depends on position and time, namely it is a function of the following form.

$$t = f(x, y, z, \tau) \tag{1}$$

The temperature field is such, the total of the temperature *t* values throughout all space, at some time τ, the expression (1) representing the equation of this field.

The temperature field may be constant (stationary or permanent) and transient (non-stationary or variable) as time τ is explicitly or nonexplicitly in equation (1) and namely:

- the temperature constant field has the equation:

$$t = f_1(x, y, z), \quad \frac{\partial t}{\partial \tau} = 0 \tag{2}$$

- the temperature transient field is expressed by equation (1)

Regarding the number of coordinates, the temperature field can be one, two or tridirectional. So, inequations (1) and (2), the temperature field is tridirectional in transient regime, respectively, constant. If the temperature is depending on two coordinates and time, the field is bidirectional transient, with the equation:

$$t = f_2(x, y, \tau), \quad \frac{\partial t}{\partial z} = 0 \tag{3}$$

and if it is expressed as a function of a coordinate and time, the field is onedirectional transient, with equation:

$$t = f_3(x, \tau), \quad \frac{\partial t}{\partial y} = \frac{\partial t}{\partial z} = 0. \tag{4}$$

The equation of temperature in the onedirectional constant field is the most simple:

$$t = f_4(x), \quad \frac{\partial t}{\partial y} = \frac{\partial t}{\partial z} = \frac{\partial t}{\partial \tau} = 0. \tag{5}$$

All points of space considered which have the same temperature t at a moment in time τ form**isothermal surface**. Because, a point in an object can not concomitant has two different temperature values, the result is that isothermal surfaces are continuous surfaces which do not intersect one another. [5]

The temperature gradient is a measurement which expresses elementary increase of temperature to a point of a temperature field, at a given time τ.

The temperature gradient is a normal vector to isothermal surface and it is numerically equal with the limit of the report between the temperature variation Δtbetween two isothermal surfaces and the distance Δn of these two, measured on normal to surface, when Δn tends to zero, namely:

$$gradt = \lim_{\Delta n \to 0} \frac{\Delta t}{\Delta n} = \frac{\partial t}{\partial n} \quad [°C/m] \tag{6}$$

The heat amount ΔQ passing through an object or from an object to another, by an isothermal surface S, in time unit $\Delta \tau$, is called **heat flow (thermal flow)**Q:

$$\dot{Q} = \frac{\Delta Q}{\Delta \tau} \quad [\text{W}], \tag{7}$$

The unitary heat flow$q_S$is the heat flow which crosses the surface unit in the time unit:

$$\dot{q}_S = \frac{\dot{Q}}{S} = \frac{\Delta Q}{S \Delta \tau} \quad [\text{W/m}^2] \tag{8}$$

where S is the area of heat exchange surface, in m^2.

Electrical analogy of heat transfer

Two systems are analogous when they have different natures, but respect similar equations with similar boundary conditions. This implies that equations describing the action of a system can be transformed into the equations of the other system by simply exchanging the variable symbols. So, Ohm's law for Electrotechnics expressing the link between DC I, voltage (potential difference) ΔU and the electrical resistance Re, has an analog form in heat transfer, by the relationship between unitary thermal flow q, temperature difference (thermal potential) Δt and a size called **thermal resistance**Rt,,namely:

$$I = \frac{\Delta U}{R_e} \ ; \quad \dot{q}_S = \frac{\Delta t}{R_t} \tag{9}$$

In equation (9), when q is measured in W/m^2 and Δt is measured in °C, the thermal resistance $Rt$is expressed in (m^2 °C)/W.

Based on this analogy, a series of concepts of DC circuits can be applied to the heat transfer problems (for example, electrical circuit has an equivalent thermal circuit and reverse) and in dynamic regime (for example, electrical modeling of some transient thermal processes). The electrical analogy of heat transfer can be used as a calculation tool and as viewing heat transfer equations by linking them to the electrical engineering domain.

Corresponding to the three fundamental types of heat transfer, we can establish the expressions for calculating the thermal resistance to conduction, convection and respectively, radiation, which can take place in the complex processes of heat exchange, equivalent electrical diagrams serially or mixed connected.

Inverse thermal resistance is called **thermal conduction**. [5]

3. THE CALCULATION METHODS FOR HEATING SYSTEM OF ELECTRICAL MACHINES

The heat flow Q (the amount of thermical energy which pass through a given surface S in time unit) expressed in J/s is numerically equal with losses of power P, expressed in W, which determine heat in any part of the machine. So, instead of heat flow, in relations used, will use losses P, directly. The object characteristics and the difference between object temperature to the environment that transfers heat determines heat transfer on the mass and on the surface of that object. For thermal calculation, the analogy is made between electrical resistance from the power of a circuit and thermal resistance R from heat flow Q.

Based on this analogy, the temperature drop in mass or surface of an object corresponding to heat flow Q, (or losses P) is determined by multiplying between thermical resistance of the object opposed to heat flow in its mass, respective its surface, R, and losses P, relation (10):

$$\theta = PR\left[{}^{\circ}\mathrm{C}\right], \qquad (10)$$

Where:$\theta = \nu_1 - \nu_2$ – is the temperature drop, over temperature or object heating, in °C.

On the relation (10) we may work with equivalent*thermal diagrams*with means which, for certain values of thermal resistance and losses, the heatθ is resulting.

The way to solve the thermal schemes, depending on the thermal resistances connection (series, parallel or mixed) is identical to the electrical schemes. [2, 3]

Next we will present the thermal resistances expressions depending on the types of heat transmission and the structure encountered in building electrical machines.

Heat transmission by radiation and convection in the case of heat transmission in axial channels and radial channels of electrical machines

In heat transmission by radiation and convection in the case of heat transmission in axial channels and radial channels of electrical machines, the transmission coefficients of heat depend on the air velocity which blows

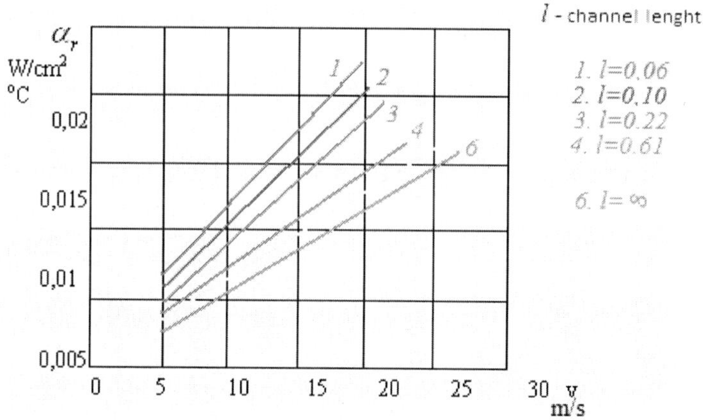

Figure 1. Transmission coefficient of heat in axial channels

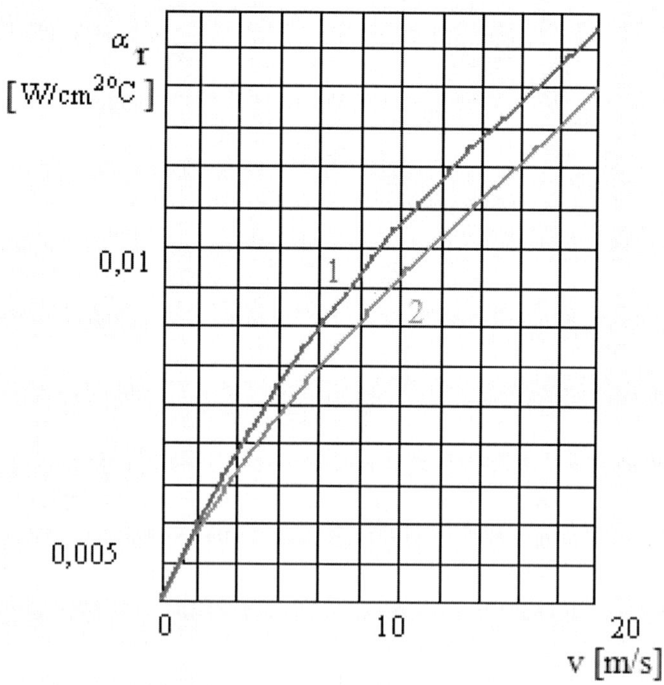

Figure 2. Transmission coefficient of heat in radial channels

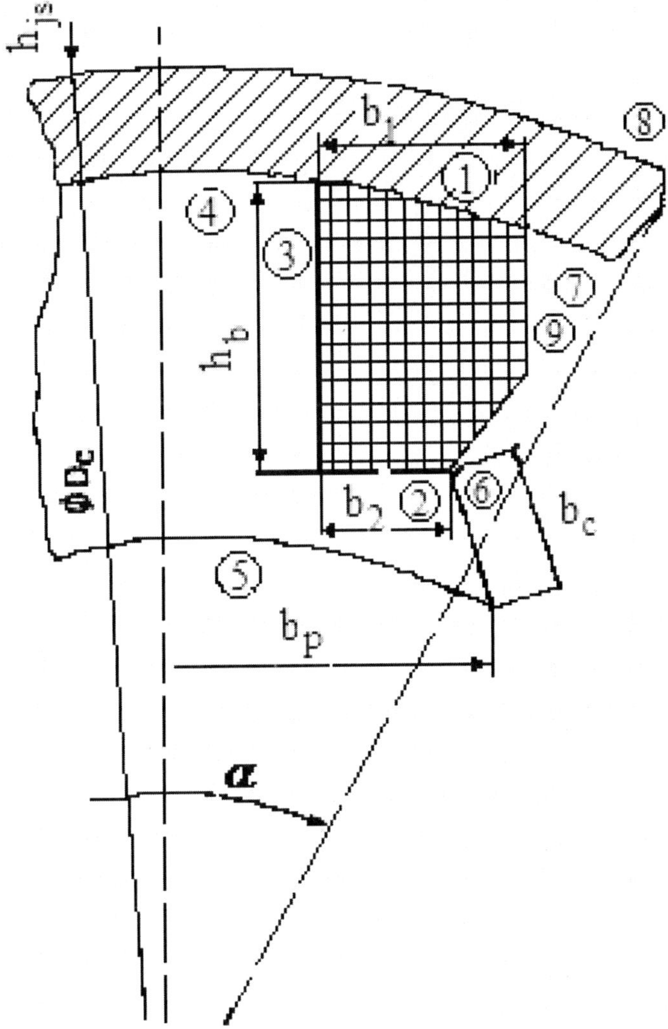

Figure 3. Transversal section through a winding pole of a DC machine.

on the surface. For certain lengths of the channel, transmission coefficients of heat will be taken from the diagrams shown in „Figure 1” for axial channels and „Figure 2” for radial channels.

4. COMMON THERMAL SCHEMES USED BY ELECTRICAL MACHINES

Operating winding of electrical machines

In "Figure 3", it is presented the transversal section through a winding pole of an electrical DC machine, for that the temperature is considered constant around the coil mass, calculating the average temperature value, the temperature drop is practically null, as well as in the coil mass and in the machine housing, for the part of housing which contributes to cooling of a magnetizing coil and corresponds an angle $\alpha = \frac{\pi}{2p}$, this is the angle between symmetry axes from a main pole and the auxiliary neighboring poles.

The real thermal diagramme of heat transmission to a coil from a pole ofDC machine is indicated in „Figure 4", where there are indicated losses by Joule effect of one coil *Pcub,* thermal resistance from coil to housing *R1* corresponding the domain ① of „Figure 3", thermal resistance from coil to pole by pole piece *R2*corresponding the domain ②of „Figure 3", thermal resistance from coil to pole by the pole body *R3* corresponding the domain ③to Figure 3, thermal resistance from pole to housing *R4* considering the pole-housing jonction $\approx$ 0,1 mm corresponding the domain ④, thermal resistance from pole to air by air-gap *R5* corresponding the domain ⑤, thermal resistance from pole to the space air between poles *R6* corresponding the domain ⑥, thermal resistance from housing to air inside machine *R7* corresponding the domain ⑦, thermal resistance from housing to environmental medium*R8* corresponding the domain ⑧, thermal resistance from coil to air *R9* corresponding the domain ⑨. All resistances are measured in ^{0}C/W.

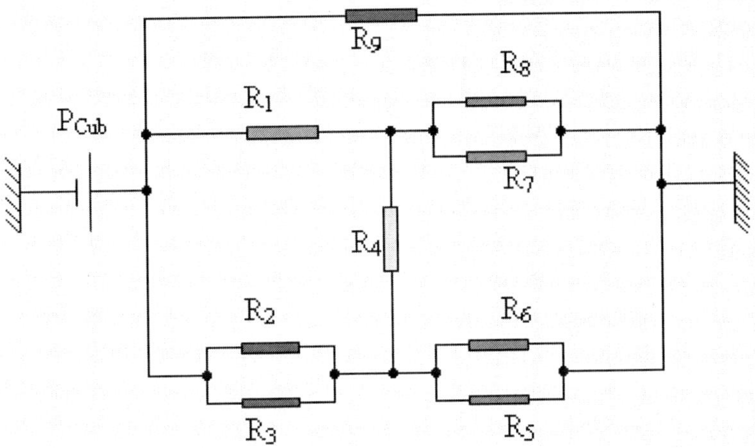

Figure 4. Main diagram of heat transmission to a coil from a pole of DC machines

For learning the total equivalent resistance we need to reproduce intermediate thermal diagrams, „Figure 5.a, b".

Thermal resistances for determining the heating of magnetizing coil from „Figure 5.a" correspond to relations:

$$R_A = \frac{R_2 R_3}{R_2 + R_3}; \; R_B = R_1; \; R_C = R_4;$$

$$R_D = \frac{R_5 R_6}{R_5 + R_6}; \; R_E = \frac{R_7 R_8}{R_7 + R_8} \qquad (11)$$

And for „Figure 5.b" after transforming the triangle *RD,RC*and *RE* from „Figure 5.a" into star *Rx, Ry*and*Rz* the relations for thermal resistances are:

$$R_x = \frac{R_C R_D}{R_C + R_D + R_E}; \; R_y = \frac{R_C R_E}{R_C + R_D + R_E}; \; R_z = \frac{R_D R_E}{R_C + R_D + R_E};$$

$$R_I = R_A + R_x; \; R_{II} = R_B + R_y \qquad (12)$$

According to (12) and the equivalent diagram from „Figure 5.b" we can determine the total equivalent resistance:

$$R_T = \frac{\left(\dfrac{R_I R_{II}}{R_I + R_{II}} + R_z\right) R_9}{\dfrac{R_I R_{II}}{R_I + R_{II}} + R_z + R_9}$$

(13)

After having known the total equivalent resistance *RT*, we can determine the heating of magnetizing coil towards the cooling air in the machine:

$$\theta_{cub} = R_T P_{cub}$$

(14)

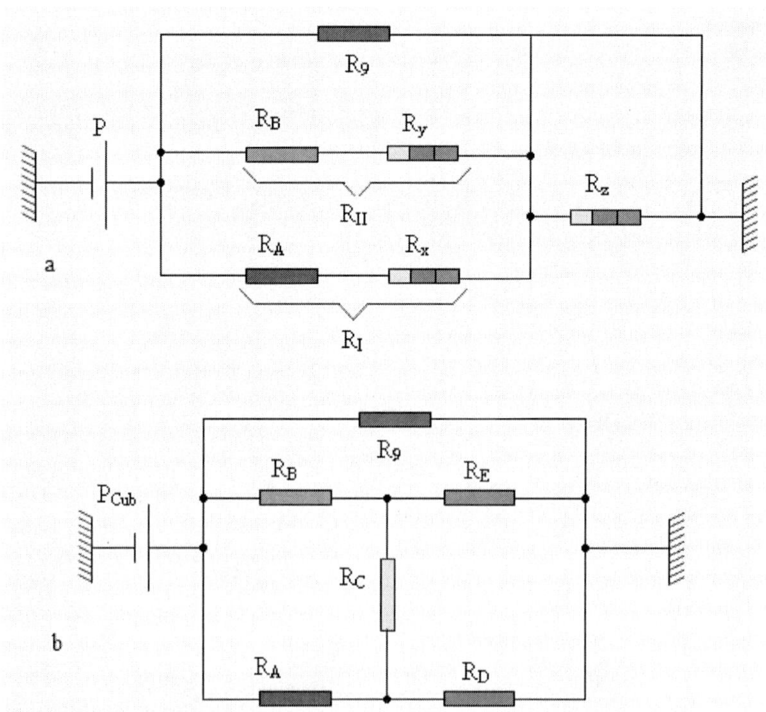

Figure 5. Intermediate thermal diagrams for determining the heating of magnetizing coil:a – real diagram; b – equivalent diagram.

The calculation of thermal resistances from the real diagram (Figure 4)

The thermal resistance from coil to machine housing

Thermal resistance from coil to machine housing *R1* is calculated according to total thickness $\beta t1$(thickness coil insulation βb, insulation thickness to the housing βc, thickness of protecting lac from impregnation $\beta l \approx 0{,}025$ cm), of equivalent thermal conductivity $\lambda 1$(thermal conductivity when coil is separately isolated λb, thermal conductibility for insulation from housing λc, thermal conductivity for protecting lac λl) and from the corresponding surface *S1* by which the heat is transmitted, considering only the side parts. [2, 3]

$$\beta_{t1} = \beta_b + \beta_c + \beta_l \quad [cm];$$

$$\lambda_1 = \frac{\beta_{t1}}{\dfrac{\beta_b}{\lambda_b} + \dfrac{\beta_c}{\lambda_c} + \dfrac{\beta_l}{\lambda_l}}$$

$$S_1 = 2b_1(l_m + b_m + 2b_1), \ [\mathrm{cm}^2] \tag{15}$$

Where:

b1 - the thickness of the coil to the upper part, in cm;

bm – the polar core width, in cm;

lm – the polar core length, in cm.

These are obtained by:

$$R_1 = \frac{\beta_{t1}}{\lambda_1 S_1} \quad [{}^{\circ}\mathrm{C/W}]. \tag{16}$$

The **thermal resistance from the coil to the polar piece.**

The thermal resistance from the coil to the polar piece, *R2*, is calculated as the thermal resistance from coil to machine housing for same insulation, but get calculating only the isolated frame bottom and the transmission surface of heat *S2*.

$$R_2 = \frac{\beta_{t2}}{\lambda_2 S_2} \quad [{}^{\circ}\mathrm{C/W}]$$

$$\beta_{t2} = \beta_{t1} \quad and \quad \lambda_2 = \lambda_1$$

$$S_2 = 2b_2\left(l_m + b_m + 2b_2\right), \ \left[\mathrm{cm}^2\right] \tag{17}$$

where *b2* is the thicknessthe coil to the below part, in cm.

The thermal resistance from the coil to the pole object

The thermal resistance from the coil to pole object R_3, is calculated according to the thickness of the insulated coil βb only when the coil is isolated separately, βm, the thickness of the insulation toward the body of the pole, the impregnation layer $\beta l \approx 0{,}025$ cm, the thermal conductivity of the insulated coil λb, thermal conductivity of insulation from the pole body λm, the thermal conduction of the protecting lac $\lambda l$and of the corresponding surface S_3.

$$\beta_{t3} = \beta_b + \beta_m + \beta_l \quad [cm]$$

$$\lambda_3 = \frac{\beta_{t3}}{\frac{\beta_b}{\lambda_b} + \frac{\beta_m}{\lambda_m} + \frac{\beta_l}{\lambda_l}} \tag{18}$$

$$S_3 = 2b_3\left(l_m + b_m + 2b_3\right), \ \left[\mathrm{cm}^2\right]$$

$$R_3 = \frac{\beta_{t3}}{\lambda_3 S_3} \quad [{}^{\circ}\mathrm{C/W}] \tag{19}$$

The thermal resistance from the pole core to the housing

The thermal resistance from the pole core to the housing R_4 is calculated according to the thickness of the parasite airgap to the pole-housing jonction at about 0,1 mm, for thermal conductivity in electrical machines whose stator is not impregnated after winding$\lambda 4$ = 0,00025 W/cm°C, or for electrical machines with the impregnated stator $\lambda 4 = \lambda l$, and, also according to the corresponding calculated surface with *bm*thepolar core width and *lm*the polar core length.

$$R_4 = \frac{0,01}{\lambda_4 S_4} \quad \left[{}^{\circ}\mathrm{C/W}\right] \tag{20}$$

The thermal resistance to the heat transmission from the core pole to airgap

Table 1. The coefficient values of heat transmission$\alpha 0$, on thenon-blast surfaces

α_0 =1,42 x 10^{-3} W/cm²°C	Iron and steel surfaces, grouted (housing surfaces)
α_0 =1,67 x 10^{-3} W/cm²°C	Iron and steel surfaces, nongrouted
α_0 =1,33 x 10^{-3} W/cm²°C	Cooper or brass surfaces, and coil impregnated or varnished surfaces

Considering the corresponding surface S_5 calculated according to the length and width of the polar piece *lp*respectively *bp*, the thermal resistance to the heat transmission from the core pole to airgap is determinate by relation:

$$R_5 = \frac{1}{\alpha_{t5} S_5} \quad \left[{}^{\circ}\mathrm{C/W}\right] \tag{21}$$

The coefficient of heat transmission on the object surface $\alpha t5$ is determined admitting that

θ'aerθaer' = 20°C is half of the total heating of the air in the machine, k = 0,8 chosen from „table 2", the coefficient of heat transmission chosen from

„table 1" is $\alpha 0 =$ 1,67 x 10^{-3}W/cm^2 ^{0}C and air velocity in the airgap $v_\delta = \frac{1}{2} v_r$.

$$\alpha_{t5} = \alpha_0 \left(1 + k\sqrt{v_\delta}\right)(1 - 0,5a) \left[\mathrm{W / cm^2 {}^\circ C}\right]$$

$$a = \frac{\theta'_{aer}}{\theta_a}$$

$a = \frac{20^o}{70^o} = 0,28$ class of insulation B, $a = \frac{20^o}{90^o} = 0,22$ = 0,22class of insulation F $S_5 = b_p l_p$ [cm^2].

Table 2. The coefficient values of blast intensities *k*

The name of the surface and the type of blast	*k*
The blast surface in the most perfect way	1,3
The surface of the front linkage of induced winding	1,0
The name of the surface and the type of blast	*k*
The surface of the active part of the induced	0,8
The surface of the magnetizing coils	0,8
The collector surface	0,6
The outer surface of the housing	0,5

The thermal resistance to heat transmission from polar piece to the space between poles

In the heat transmission to space between poles, by polar pieces, the thermal resistance is determined regarding the coefficient of heat transmission $\alpha t6$ and the surface located on the polar width of the object *bc lm* polar core length:

$$R_6 = \frac{1}{\alpha_{t6} S_6} \quad [°C/W]$$

$$S_6 = 2 b_c l_m \left[cm^2\right]$$

$$\alpha_{t6} = 1{,}6.7 \times 10^{-3}\left(1 + 0{,}8\sqrt{v_6}\right)\left(1 - 0{,}5a\right) \left[W/cm^2 °C\right], \; cu v_6 = v_p \tag{22}$$

The thermal resistance to heat transmission from the housing to interior

When heat transmission is done from housing to interior, the thermal resistance R_7 is determined the same as in heat transmission to the space between poles, by polar pieces with $\alpha t7 \approx \alpha t6$, $v_7 = v_6$, and the corresponding surface S_7 is calculated regarding the interior diameter of the housing *Dc*, the number of pole pairs *p* and the housing length *lc*.

$$R_7 = \frac{1}{\alpha_{t7} S_7} \quad [°C/W]$$

$$S_7 \approx \frac{\pi D_c}{4p} l_c - \left(b_m + 2b_1\right)\left(l_m + 2b_1\right) \quad \left[cm^2\right] \tag{23}$$

The thermal resistance to heat transmission from housing to exterior (quiet air)

The heat transfer from the housing to outer where the air is quiet, thermal resistance is calculated taking into account the heat transmission coefficient $\alpha 0 = 1{,}42 \times 10^{-3}\ W/cm^{2}\,{}^0C$ determined in „table 1", by the appropriate surface S_8, at which involved, the height of stator yoke *hjs*, the inner diameter of the housing *Dc*, the number of pole pairs *p* and the housing length *lc*.

$$R_8 = \frac{1}{\alpha_0 S_8} \quad [°C/W]$$

$$S_8 = \frac{\pi\left(D_c + 2h_{js}\right)}{4p} l_c \tag{24}$$

The thermal resistance to heat transmission from coil directly to cooling air in the machine

When the heat transmission is done from coil directly to cooling air in the machine, the thermal resistance depends on the thickness and the thermal conductivity of coil insulation when the coil is isolated separately βbandλb, by the coefficient of heat transmission $\alpha t9$, and by the corresponding surface S_9.

$$R_9 = \frac{\beta_b}{\lambda_9 S_9} + \frac{1}{\alpha_{t9} S_9} \quad \left[{}^{o}C / W\right]$$

$$\alpha_{t9} = \alpha_0 \left(1 + 0,8\sqrt{v_9}\right)\left(1 - 0,5a\right) \left[W / cm^2 {}^{\circ}C\right],$$

$$\alpha_0 = 1,33 \bullet 10^{-3}.\ W / cm^2 {}^{\circ}C, \text{ from table 1; } v_9 = v_6;$$

$$S_9 \approx 2\left(b_m + l_m + 4b_1\right)1,1h_b \quad \left[cm^2\right] \tag{25}$$

5. THE LOSSES BY JOULE EFFECT, A SINGLE COIL

The losses by Joule effect *Pcub* are the total losses in the magnetizing reeling *Pcuc* reported to the number of poles of the electrical machine *2p*.

$$P_{cub} = \frac{P_{cuc}}{2p} \quad [W] \tag{26}$$

6. THE TOTAL HEAT OF MAGNETIZING REELING

Heating the magnetizing coil to the air in the machine, is determined by the relation (14) together with heating the air to the contact with the coil ($\theta_{aer} \approx 20°C$ – from rotor to stator) determining heating of the magnetizing reeling towards the external environment.

$$\theta_{\text{cue}} = \theta_{\text{cub}} + \theta'_{aer} < \theta_{admis} \quad (27)$$

7. THE DIAGRAMS OF VENTILATION AND CALCULATION OF AERODYNAMIC RESISTANCES

In the construction of rotating electrical machines, the cooling air is most frequently used. Depending on the construction type and the type of protection of the electrical machine the best ventilation system is chosen.[1]

Calculating ventilation of rotating electrical machine is done as follows:

- establishing the ventilation diagram andcalculating aerodynamic resistances;
- calculating the pressure determined by means of air entrainment;
- calculating air flow at different sections.

Preparing the ventilation diagram for a rotating electrical machine requires prior execution of the diagram to the scale of general assembly (longitudinal section and cross section) and establishing the routes where cooling air circulates. For calculation aerodynamic resistances we need to know longitudinal and transversal dimensions of all pipes that form the routes for air circulation.

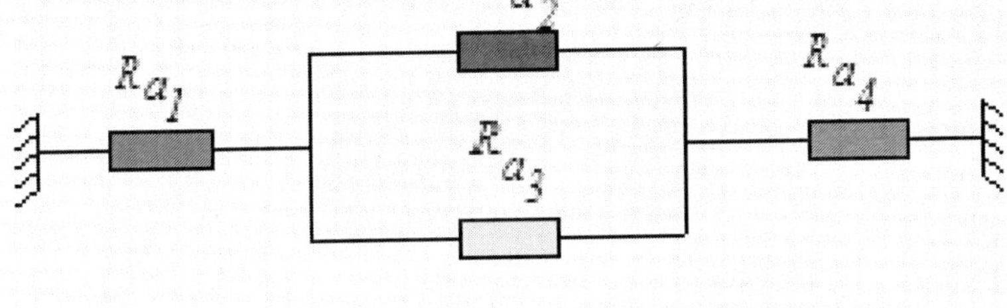

Figure 6. The diagram of mixed connection of aerodynamic resistances

When we solve an aerodynamic diagram, we aim to find the equivalent resistance *Rae*.. Between aerodynamic diagrams used for air circuits and electrical diagrams may be an analogy, the aerodynamic resistances act as passive elements.

Determining the equivalent resistance depends on how aerodynamic resistances are connected:

- to series arrangement *Rae* is the sum of the aerodynamic resistances and it is determined by relation:

$$R_{ae} = \sum_{i=1}^{n} R_{ai} \tag{28}$$

- to shunt connection*Rae* is determined by relation:

$$\frac{1}{\sqrt{R_{ae}}} = \sum_{i=1}^{n} \frac{1}{\sqrt{R_{ai}}} \tag{29}$$

- to mixed connection, „Figure 6", *Rae* is calculated according to relations of calculation for the series arrangement and shunt connection and we obtain:

$$R_{ae} = R_{a1} + \frac{R_{a2} R_{a3}}{\left(\sqrt{R_{a2}} + \sqrt{R_{a3}}\right)} + R_{a4} \tag{30}$$

Bilateral axial-radial ventilation for an electrical machine in protected construction

In „Figure 7.a" are represented the longitudinal section and the frontal view for an induction motor with the rotor in short-circuit, in protected construction, and with arrows we indicate routes that circulate the cooling air.

Typically, in asynchronous machine the role of inductor is accomplished by the stator, while the rotor is the induced machine. The machine excitation is done in AC. The collector body is of „collector rings" type.

The asynchronous machine can not develop a couple unless the speed *„n"* of the rotor is different by the speed *„n1"* ofthe stator rotating field. [2, 3]

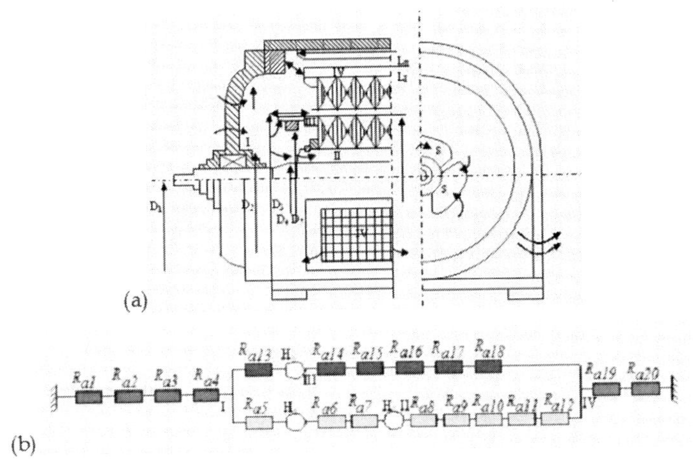

Figure 7. Ventilation diagram of asynchronous electrical machine with bilateral radial ventilation: a) physical diagram; b) main diagram.

Because depression created in the rotor bars ends and the rotor ribs (the electrical machine has not its own ventilator), the air is absorbed through the windows of the two fore-shields, washes the active elements and it is thrown out through the windows of its middle, by the bottom of the housing.

Air flow is symmetrical to the middle of the machine length, and is, therefore, sufficient to calculate ventilation only on one side, which is half of the machine.

The sources of pressure are denoted by „*H*", the aerodynamic resistance by „*Ra*" and are divided in sections. [3, 11, 12]

8. SECTION I

We established I section, which contains aerodynamic resistances $Ra1$, Ra_2, Ra_3, andRa_4, which are connected in series and is calculated the total aerodynamic resistance for section I, according the relation:

$$R_{atrI} = \sum_{i=1}^{4} R_{ai} \tag{31}$$

For determining total aerodynamic resistance of the I section we calculate, in turns, each aerodynamic resistance which is part of the I section regarding the coefficients of aerodynamic resistances and the sections of the channels by which the air circulates.

Table 3. The values of the resistance coefficient to the air admission pipes

Conditions of entry	ζ
Entry with sharp edges	62×10^{-2}
Entry with edges at angle of 90°	30×10^{-2}
Entry with rounded edges	$(12{,}5\ldots0) \times 10^{-2}$

Table 4. The values of the resistance coefficient to deviation in different angles

The angle of deviation	ζ
135°	32×10^{-2}
90°	70×10^{-2}

The air enters through the shield windows under sharp edges (blinds) by narrowing from the external environment to the input section. From „table 3", for sharp edges $\zeta_{as} = 62 \cdot 10^{-2}$, and for narrowing $\frac{S_1}{S_{ext}} \approx 0$, as in „Figure 8"

8" $\zeta_{ing} \approx 32 \cdot 10^{-2}$.

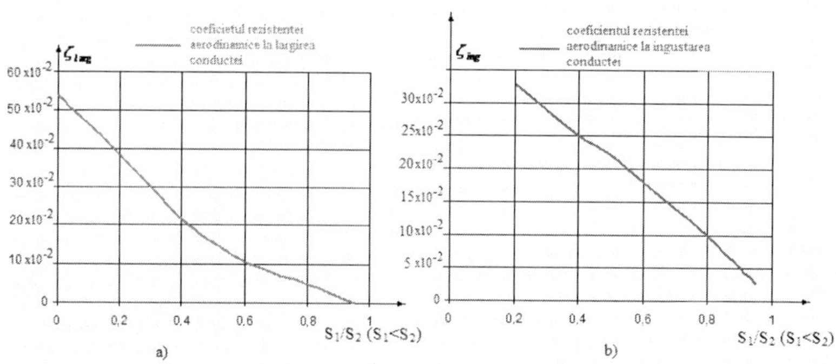

Figure 8. The aerodynamic coefficient resistance to widening and narrowing core

Aerodynamic resistance *Ra1* is calculated according to the net input surface in the shield windows S_1, to the coefficient of aerodynamic resistance to narrowing *ζing* and to the coefficient for sharp edges *ζas.*

$$R_{a1} = \frac{\zeta_{as} + \zeta_{ing}}{S_1^2} = \frac{(62 + 32) \cdot 10^{-2}}{S_1^2} \tag{32}$$

Where - $S_1 = k_i S_i$ $\left[m^2\right]$

- *Si* is the total surface ofthe shield windows, in m^2;
- *ki* is the coefficient of reduction in section because the coatings required by the machine protection: ki=0,6÷0,7

There is a relaxation of the air coming out of the shield and to the input in the machine and a deviation to 135° by deflector (Figure 7.a)

$$S_2 = \frac{\pi}{4}\left(D_1^2 - D_2^2\right) \quad \left[m^2\right] \tag{33}$$

For

$\frac{S_1}{S_2} = \ldots \Rightarrow \zeta_{larg} = \ldots.$, and for the deviation to 135° from „table 4" $\Rightarrow \zeta r = 32 \cdot 10^{-2}$

The aerodynamic resistance *Ra2* is calculates according to the net output surface of windows shield *S1*, to the coefficient of aerodynamic resistance to enlargement *ζlarg* and to the coefficient to the air relaxation of shield output *ζas.*

$$R_{a2} = \frac{\zeta_{larg} + \zeta_r}{S_1^2} \tag{34}$$

Next, there is a relaxation to the output of deflector under sharp edges.

$$R_{a2} = \frac{\zeta_{larg} + \zeta_r}{S_1^2} \tag{35}$$

For

$\frac{S_2}{S_3} = \ldots \Rightarrow \zeta_{larg} = \ldots.,$, and for sharp edges $\zeta_{as} = 62 \cdot 10^{-2}$

We calculate the aerodynamic resistance *Ra3* regarding the net input surface in the machine *S2*, the coefficient of aerodynamic resistance to widening *ζlarg* and the coefficient for sharp edges*ζas*.

$$R_{a3} = \frac{\zeta_{larg} + \zeta_{as}}{S_2^2} \tag{36}$$

At the rotor winding input occurs a narrowing and there are sharp edges for the short-circuited rotor.

$$S_4 = \frac{\pi}{4}\left(D_5^2 - D_4^2\right) \ \left[m^2\right] \tag{37}$$

For

$\frac{S_4}{S_3} = \ldots \Rightarrow \zeta_{ing} = \ldots.,$, and for sharp edges $\zeta_{as} = 62 \cdot 10^{-2}$.

We calculate the aerodynamic resistance Ra_4regarding the net input surface under rotor winding *S4*, the coefficient of aerodynamic resistance to narrowing *ζing*and the coefficient for sharp edges *ζas*.

$$R_{a4} = \frac{\zeta_{ing} + \zeta_{as}}{S_4^2} \tag{38}$$

After we calculated the aerodynamic resistances *Ra1, Ra2, Ra3, Ra4,* regarding the narrowing or widening ventilation channels by relation (31) we calculate the *total aerodynamic resistance of the section I.*

9. SECTION II

The II section contains aerodynamic resistances *Ra5, Ra6, Ra7, Ra8, Ra9, Ra10, Ra11, Ra12,* connected in series and we calculate the total aerodynamic resistance to the section II with the relation:

$$R_{atrII} = \sum_{i=5}^{12} R_{ai} \quad (39)$$

To calculate the aerodynamic total resistance by section II, we calculate, as for section I, each aerodynamic resistance, in turns, considering the widening and the narrowing of the ventilation channels.

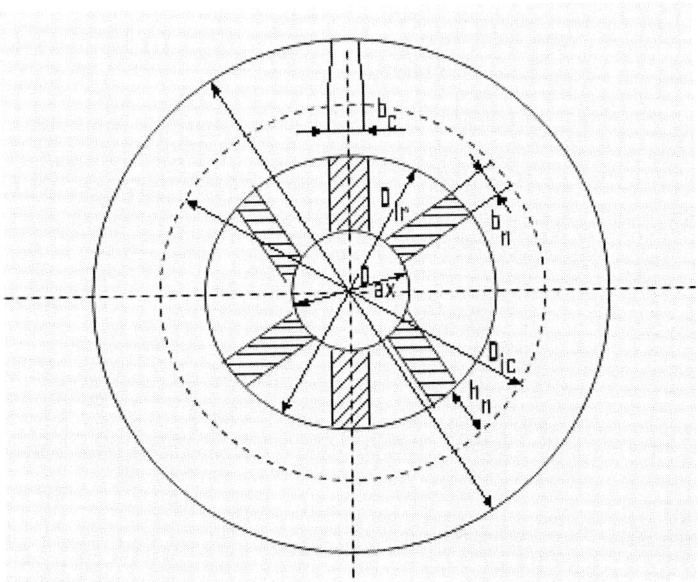

Figure 9. Cross section in the rotor with ribbed shaft

When the air begins to cross the section II, it enters under the rotor core which occurs in narrowing, and the input is done under sharp edges (Figure 9).

$$S_5 = \frac{\pi}{4}\left(D_{ir}^2 - D_{ax}^2\right) - n_n b_n h_n \quad \left[m^2\right] \quad (40)$$

Where:

- *Dir*, is the interior diameter of the armature core disc, in m;
- *Dax* is the diameter of the shaft, in m;

- *bn* is the ridge width, in m;
- $h_n = \frac{D_{tr} - D_{ax}}{2}$ is the ridge height, in m;
- *nn* is the ridges number.

For $\frac{S_5}{S_4} = ... \Rightarrow \zeta_{ing} = ...$ and for sharp edges $\zeta_{as} = 62 \cdot 10^{-2}$.

We calculate the aerodynamic resistance *Ra5*, regarding the net surface to the output under the rotor core *S5*, the coefficient of aerodynamic resistance to the narrowing *ζing* and the coefficient for sharp edges *ζas*.

$$R_{a5} = \frac{\zeta_{ing} + \zeta_{as}}{S_5^2} \tag{41}$$

For the input of radial ventilation channels of the rotor under sharp edges occurs a deviation $90^\circ, \zeta_c = 70 \cdot 10^{-2}$, from „table 3" $3'' \left(\zeta_{as} = 62 \cdot 10^{-2}\right)$ and a narrowing.

$$S_6 = 0{,}9\pi D_{ir} b_v \frac{n_v}{2} \quad \left[m^2\right] \tag{42}$$

Where:

- *bv*is the width of the radial channel for ventilation, in m;
- *nv* is the number of radial channels for ventilation;
- 0,9 is the coefficient which takes into consideration the presence of spacers.

For $\frac{S_6}{S_5} = ... \Rightarrow \zeta_{ing} = ...$

The aerodynamic resistance *Ra6* is calculated, using the relation (42) which takes into consideration the net surface to the input of radial channels of the ventilator *S6*, the coefficient of aerodynamic resistance to narrowing *ζing* and the coefficient for sharp edges *ζas*.

After the air entered in the channels of ventilation the relaxation occurs (Figure 9)

$$S_7 = 0{,}9\pi D_{ic} b_v \frac{n_v}{2} \quad \left[m^2\right] \tag{43}$$

Where:

- D_{ic} is the diameter of the rotor slots base, in m.

For $\frac{S_6}{S_7} = \ldots \Rightarrow \zeta_{l\arg} = \ldots.$

The aerodynamic resistance *Ra7* is calculated, taking into consideration the relaxation in the channels of ventilation:

$$R_{a7} = \frac{\zeta_{l\arg}}{S_6^2} \tag{44}$$

A new narrowing occurs to the entry in the region of rotor coils under sharp edges, for the short-circuited rotor or the rounded edges of rotors and widening to the outside rotor.

$$S_8^{'} = S_7 - z_2 b_{c2} b_v \frac{n_v}{2} \quad \left[m^2\right] \tag{45}$$

$$S_8^{''} = 0{,}9\pi D_r b_v \frac{n_v}{2} - Z_2 b_{c2} b_v \frac{n_v}{2} \quad \left[m^2\right] \tag{46}$$

Where:

- D_r is the exterior diameter of the rotor, in m;
- z_2 is the number of rotor slots;
- b_{c2} is the width of the rotor slot, in m.

For

$\frac{S_8^{'}}{S_7} = ... \Rightarrow \zeta_{ing} = ...$ and for sharp edges $\zeta_{as} = 62 \cdot 10^{-2}$.

For $\frac{S_8^{'}}{S_7} = ... \Rightarrow \zeta_{ing} = ...$

We obtained the aerodynamic resistance *Ra8*, calculated to the input of the coils region by narrowing the ventilation channel and the output of rotor by widening the ventilation channels for short-circuited rotor where the edges are sharp:

$$R_{a8} = \frac{\zeta_{ing} + \zeta_{as} + \zeta_{larg}}{S_8^{'2}} \tag{47}$$

And *Ra8*, for phase-wound rotor where the resistance of coils rounded edges are neglected is:

$$R_{a8} = \frac{\zeta_{ing} + \zeta_{larg}}{S_8^{'2}} \tag{48}$$

At the transition from the rotor coils zone to the stator coils zone and the input under rounded edges a widening (or a narrowing) occurs $\left(\zeta_r = 12{,}5 \cdot 10^{-2}\right)$.

$$S_9^{'} = 0{,}9\pi D b_v \frac{n_v}{2} - z_1 b_{c1} b_v \frac{n_v}{2} \quad \left[m^2\right] \tag{49}$$

Where:

- D is interior diameter of the stator, in m;
- z_1 is the number of slots of the stator;
- b_{c1} is the width of stator slots, in m.

For the rapport $\frac{S_8^{''}}{S_9^{'}}$ or $\frac{S_9^{'}}{S_8^{''}}$ (the lower section is used to the numerator) we determine *ζlarg* or *ζing*. For determining the resistance we put at the denominator the smallest section.

The resistance *Ra9* is obtained, taking into account the net surface in the stator coils zone *S9*, the aerodynamic resistance coefficient to narrowing/widening *ζing g*and the coefficient for the input under sharp edges *ζr*.

$$R_{a9} = \frac{\zeta_{l\arg(ing)} + \zeta_r}{S_{8(9)}^{''2}} \tag{50}$$

In the stator coils zone and to the output between coils, the net surface increases and a relaxation occur.

$$S_{10} = 0,9\pi\left(D + 2h_{cl}\right)b_v \frac{n_v}{2} \quad \left[m^2\right] \tag{51}$$

$$S_9^{'} = S_{10} - z_1 b_{c1} b_v \frac{n_v}{2} \quad \left[m^2\right] \tag{52}$$

Where:

- *hc1* is the height of the stator slot, in m.

For $\frac{S_9^{'}}{S_9^{''}} = ... \Rightarrow \zeta_{l\arg}^{'} = ...$; For $\frac{S_9^{''}}{S_{10}} = ... \Rightarrow \zeta_{l\arg}^{''} = ...$

The resistance *Ra10*, is calculated taking into account the net surface in the stator coils zone *S9* and the net surface to the output between coils *S10*, the aerodynamic resistance coefficient to widening *ζlarg* in both zones.

$$R_{a10} = \frac{\zeta_{l\arg}^{'}}{S_9^{'2}} + \frac{\zeta_{l\arg}^{''}}{S_9^{''2}} \tag{53}$$

To the exterior armature core disks in the radial channels of the stator a widening is produced, again, the surface *S11*is bigger than the surface *S10*.

$$S_{11} = 0,9\pi D_e b_v \frac{n_v}{2} \quad \left[m^2\right] \tag{54}$$

Where:

- *De*is the exterior diameter of the stator core disk, in m.

For $\frac{S_{10}}{S_{11}} = ... \Rightarrow \zeta_{l\arg} = ...$

The result is R_{a11}:

$$R_{a11} = \frac{\zeta_{l\arg}}{S_{10}^2} \tag{55}$$

To the output of radial channels of the stator, the air suffers a relaxation again.

$$S_{12} = \pi D_c \frac{L_1}{2} \quad \left[m^2\right] \tag{56}$$

Where:

- *L1*is the geometric length of the machine plus the thickness pressure rings, in m.

For $\frac{S_{11}}{S_{12}} = ... \Rightarrow \zeta_{l\arg} = ...$

Because there is a widening of ventilation channels to the output of the stator radial channels *R a12*is calculated regarding the aerodynamic resistance coefficient to widening and S_{11}:

$$R_{a12} = \frac{\zeta_{l\arg}}{S_{11}^2} \tag{57}$$

10. SECTION III

The aerodynamic resistances *Ra13, Ra14, Ra15, Ra16, Ra17, Ra18*constitute the section III, they are serially connected. Taking into account the equation (28) the total aerodynamic resistance is calculated by the section III with equation:

$$R_{atrIII} = \sum_{i=13}^{18} R_{ai} \tag{58}$$

After the air got out of the stator radial channels, it will cross the rotor ends coils by deviation to 135°for which

ζc=32·10−2

(Figure 7 and Figure 10).

$$S_{13} = \pi D_3 l_{br} \quad \left[m^2\right] \tag{59}$$

Where:

- *lbr* is the axial length of the rotor coil end, in m.

Because there is no widening or narrowing, the resistance *R a13*is calculated regarding the crossing surface to the rotor coil ends and the coefficient for rounded edges:

$$R_{a13} = \frac{\zeta_c}{S_{13}^2} \tag{60}$$

After that a narrowing to input follows between rotor bars ends and sharp edges. *ζas*= 62 x 10^{-2} for short-circuited rotors from bars with sharp edges. For the rotor with bars with rounded edges *ζas*=0.

$$S_{14} = \pi D_3 \left(l_{br} - l_{bi}\right) - z_2 b_{r2} \left(l_{br} - l_{bi}\right) \quad \left[m^2\right] \tag{61}$$

Where:

- *lbi* is the shorting ring width, in m(Figure 10)

For $\frac{S_{14}}{S_{13}} = \ldots \Rightarrow \zeta_{ing} = \ldots.$

Resulting the resistance *Ra13* calculated using the net surface to the input between the rotor bars ends and the coefficients of aerodynamic resistances to narrowing under sharp edges:

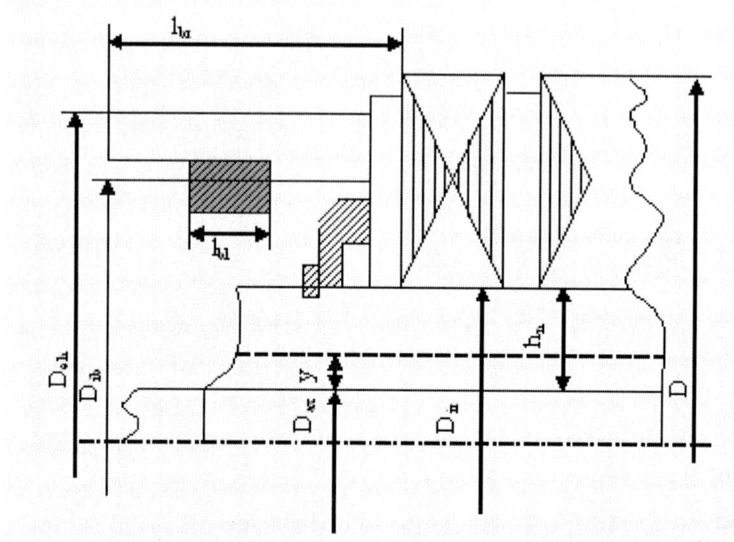

Figure 10. Partial longitudinal section through a rotor cage induction motor with ribbed shafts and bars ends having a ventilator effect (the rotor from Figure 7)

$$R_{a14} = \frac{\zeta_{as} + \zeta_{ing}}{S_{14}^2} \quad or \quad R_{a14} = \frac{\zeta_{ing}}{S_{14}^2} \tag{62}$$

The isolated support of the bandages does not allow air to enter through the coils and as a result in the right front end, to winding motors only a narrowing occurs.

To the ends output (the bars) of the rotor coils there is a relaxation of the air, a relaxation caused by widening the channel (Figure 7)

$$S_{15} = \pi D l_d \quad \left[m^2\right] \tag{63}$$

For $\frac{S_{14}}{S_{15}} = \ldots \Rightarrow \zeta_{larg} = \ldots.$

Is obtained:

$$R_{a15} = \frac{\zeta_{ing}}{S_{14}^2} \tag{64}$$

There is a passing through the spaces between the stator coils ends which can be of two types.

- prefabricated coils
- winding to the round conductor.

The prefabricated coils (from profiled conductor or bars) are represented in „Figure 11".

We determine the surface *S16* under the front part to which we consider the narrowing of the lower layer.

$$S_{16} = \pi D l_f \quad \left[m^2\right] \tag{65}$$

We admit then, that the distances *d1*and *k* between coil ends, to lower part, are equal to those at the upper part (*d1* and *k* are constant on front ends height).

Narrowing at the lower layer (Figure 11)

$$S_{16}^{'} = \left[l_{bs}d_1 + \left(l_{mi} - b_c\right)a + kc - b_d d_1 n_d\right]z \quad \left[m^2\right] \tag{66}$$

Where:

- *nd* is the number of distances between coil adjacent ends;
- *lmi* is the dental corresponding to the mean diameter of the coil lower side, in m.

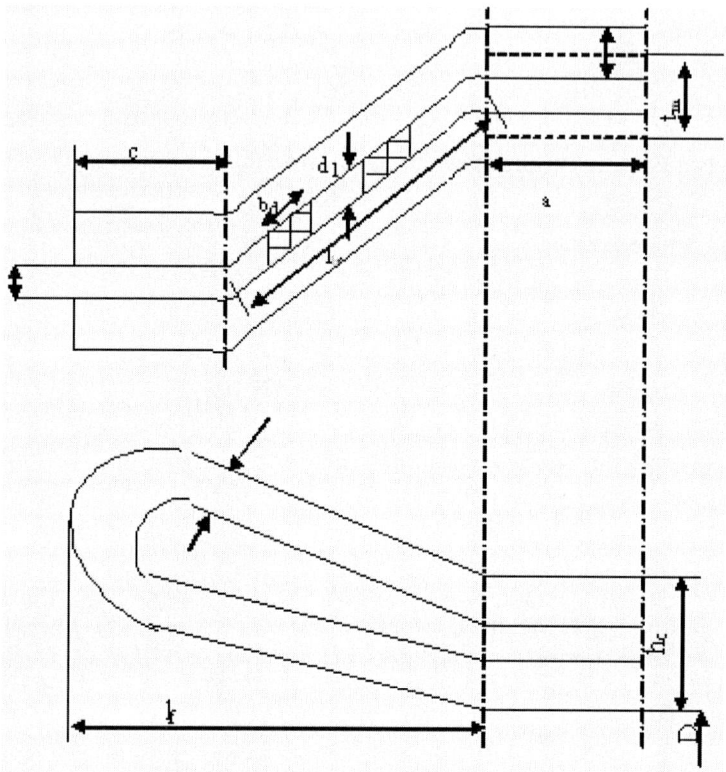

Figure 11. End of prefabricated stator coil

For

$\frac{S_{16}^{'}}{S_{16}} = \ldots \Rightarrow \zeta_{ing} = \ldots.$, and for rounded edges $\zeta_r = 12{,}5 \times 10^{-2}$

The resistance $R'_{a16}$is:

$$R_{a16}^{'} = \frac{\zeta_r + \zeta_{ing}}{S_{16}^{'2}} \quad (67)$$

For the output relaxation from lower layer S''_{16} is:

$$S_{16}^{''} \approx \pi\left(D + h_{c1}\right) l_f \quad \left[m^2\right] \quad (68)$$

Where:

hc1 is height of stator slot, in m.

For $\frac{S'_{16}}{S''_{16}} = ... \Rightarrow \zeta_{larg} = ...$

We obtained:

$$R''_{a16} = \frac{\zeta_{larg}}{S'^{2}_{16}} \tag{69}$$

Narrowing in the upper layer and the input under the slightly rounded edges.

$$S'''_{16} \approx S'_{16}; \quad \text{for } \frac{S'_{16}}{S''_{16}} = ... \Rightarrow \zeta_{ing} = ...$$

$$S'''_{16} \approx S'_{16}; \quad \text{for } \frac{S'_{16}}{S''_{16}} = ... \Rightarrow \zeta_{ing} = ...$$

For

For the slightly rounded edges from „table 3" ζr= 12,5 x 10^{-2}

The resistance $R'''a16$ is:

$$R'''_{a16} = \frac{\zeta_{r} + \zeta_{ing}}{S'^{2}_{16}} \tag{70}$$

The total resistance to airflow between the ends of stator winding is the sum of calculated resistances for narrowing in the lower layer, the relaxation to the output from lower layer and the narrowing in the upper layer and the input under the slightly rounded edges:

$$R_{a16} = R'_{a16} + R''_{a16} + R'''_{a16} \tag{71}$$

In the case of winding from round conductor (from wire), the air can not penetrate through the coil ends, because spaces between coils are obstructed of isolated strips between phases, so the air avoid them. The distance X between the end of stator coil and deflector, must be large enough not to exclude the air (Figure 7.a)

The narrowing in the section X.

$$S_{16} \cong \pi\left(D+h_{c1}\right)X\left[m^2\right] \tag{72}$$

For $\frac{S_{16}}{S_{15}} = \ldots \Rightarrow \zeta_{ing} = \ldots.$ and the resistance *Ra16* is:

$$R_{a16} = \frac{\zeta_{l\arg}}{S_{16}^2} \tag{73}$$

After passing by front ends of stator winding there is a relaxation of air after widening the ventilation channel.

$$S_{17} \approx \pi\left(D+2h_{c1}\right)\left(l_f + X\right) \quad \left[m^2\right] \tag{74}$$

For $\frac{S_{16}}{S_{17}} = \ldots \Rightarrow \zeta_{ing} = \ldots.$ and aerodynamic resistance *Ra17* is:

$$R_{a17} = \frac{\zeta_{l\arg}}{S_{16}^2} \tag{75}$$

The air flows in a narrow channel between housing ribs to a deviation of 135° (ζc= 62 x 10^{-2}).

$$S_{18} = \pi D_c p - b_{nc} p n_{nc} \quad \left[m^2\right] \tag{76}$$

Where:

- *bnc* is the housing rib width, in m;
- *nnc* is the number of the housing ribs;
- *Dc, p* (from Figure 7.a) in m.

For $\frac{S_{18}}{S_{17}} = \ldots \Rightarrow \zeta_{ing} = \ldots.$ and so:

$$R_{a18} = \frac{\zeta_c + \zeta_{as} + \zeta_{ing}}{S_{18}^2} \tag{77}$$

11. SECTION IV

The total resistance of section IV is calculated as in the other sections with the equation (26) for the series connection:

$$R_{atrIV} = \sum_{i=19}^{20} R_{ai} \tag{78}$$

A deviation at 90° occurs ($\zeta c = 70 \times 10^{-2}$) to the output sided windows (Figure 7.a) without the narrowing or widening of ventilation channels:

$$S_{19} = \frac{L_n B_n}{2} \left[m^2\right] \tag{79}$$

And the result is:

$$R_{a19} = \frac{\zeta_c}{S_{19}^2} \tag{80}$$

The output of the air from the housing is done to the deviation at 90°($\zeta c = 70 \times 10^{-2}$) for the relaxation of the outside air (Figure 7.a).

$$S_{20} = \frac{1}{2} L_p h_p k_i \quad \left[m^2\right] \tag{81}$$

For this *ki* we must take into account the calculation of*Ra1*.

For $\frac{S_{20}}{S_{ex}} \approx 0 \Rightarrow \zeta_{larg} = 60 \cdot 10^{-2}$ and so:

$$R_{a20} = \frac{\zeta_c + \zeta_{l\arg}}{S_{20}^2} \qquad (82)$$

In the scheme from „Figure 7.b", *Hb* is the pressure produced by the ends of moving rotor bars, *Hn*is the pressure produced by rotor ribs, and *Hd* is the pressure produced by rotor spacers.

12. THE CALCULATION AIR FLOW

The calculation of air flow is done after we solved the ventilation scheme regarding the air flows of different sections and after we calculated the aerodynamic resistances values and pressures values created by sources.

Bilateral radial-axial ventilation for an electrical machine in protected construction

We calculate the flows for the asynchronous electrical machine with bilateral radial ventilation from „Figure 7.a", the cooling air is driven only by the effect of ventilator of ends of rotor bars and of ribs and spacers on rotor.

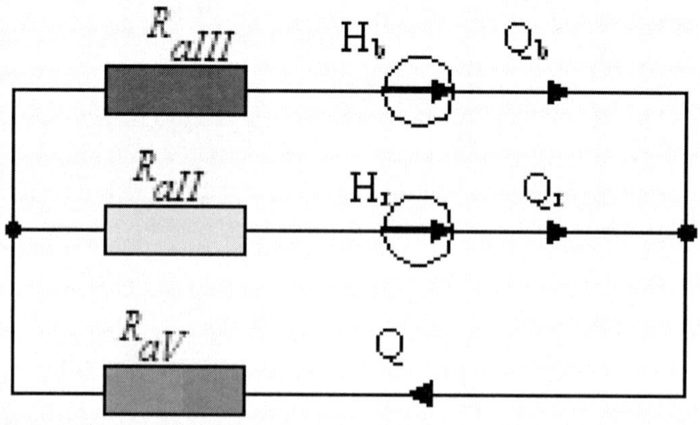

Figure 12. Variation for the representation of scheme from Figure 7.b

The flows *Qb, Qr*and *Q* by sections III, II, and V, respectively I and IV from „Figure 7.b", are determined by equations:

$$Q_b = \sqrt{y} \quad \left[m^3/s\right] \tag{83}$$

$$Q_r = \frac{1}{\sqrt{R_{aII}}}\sqrt{R_{aIII}Q_b^2 - H_b + H_r} \quad \left[m^3/s\right] \tag{84}$$

$$Q = Q_b + Q_r \tag{85}$$

When the flow which enters and exit from the machine is less than the flow calculated for half of the machine it is necessary to adopt a solution by decreasing the equivalent resistances on the sections by sections increasing, or giving up the constructive solution adopted and finding other solutions.

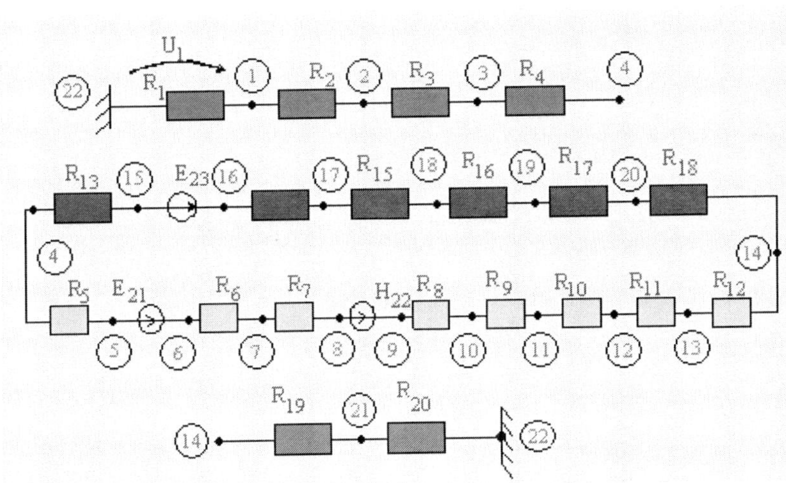

Figure 13. The principal ventilation scheme of asynchronous electrical machine with bilateral radial ventilation

The example of calculation for asynchronous electrical machine with bilateral radial ventilation:

- We calculate the aerodynamic resistance on every section, starting from the nominal data.
- Equivalent resistance by section I is: $R_{atrI} = \sum_{i=1}^{4} R_{ai}$

- Equivalent resistance by section II is: $R_{atrII} = \sum_{i=5}^{12} R_{ai}$
- Equivalent resistance by section III is: $R_{atrIII} = \sum_{i=13}^{18} R_{ai}$
- Equivalent resistance by section IV is: $R_{atrIV} = \sum_{i=19}^{20} R_{ai}$

By means of the computer program PANCIA (Program Analysis of Analog Circuits) we obtain the system solution. [7, 8, 9, 10]

• The unknowns of the system								
	V1	V2	V3	V4	V5	V6	V7	V8
V9	V10	V11	V12	V13	V14	V15	V16	V17
V18	V19	V20	V21	I21	I22	I23		

- The system of equations

+ (+G1+G2)*V1 + (-G2)*V2 = 0
+ (-G2)*V1 + (+G2+G3)*V2 + (-G3)*V3 = 0
+ (-G3)*V2 + (+G3+G4)*V3 + (-G4)*V4 = 0
+ (-G4)*V3 + (+G4+G5+G13)*V4 + (-G5)*V5 + (-G13)*V15 = 0
+ (-G5)*V4 + (+G5)*V5 + (+1)*I21 = 0
+ (+G6)*V6 + (-G6)*V7 + (-1)*I21 = 0
+ (-G6)*V6 + (+G6+G7)*V7 + (-G7)*V8 = 0
+ (-G7)*V7 + (+G7)*V8 + (+1)*I22 = 0
+ (+G8)*V9 + (-G8)*V10 + (-1)*I22 = 0
+ (-G8)*V9 + (+G8+G9)*V10 + (-G9)*V11 = 0
+ (-G9)*V10 + (+G9+G10)*V11 + (-G10)*V12 = 0
+ (-G10)*V11 + (+G10+G11)*V12 + (-G11)*V13 = 0
+ (-G11)*V12 + (+G11+G12)*V13 + (-G12)*V14 = 0
+ (-G12)*V13 + (+G12+G18+G19)*V14 + (-G18)*V20 + (-G19)*V21 = 0
+ (-G13)*V4 + (+G13)*V15 + (+1)*I23 = 0
+ (+G14)*V16 + (-G14)*V17 + (-1)*I23 = 0
+ (-G14)*V16 + (+G14+G15)*V17 + (-G15)*V18 = 0
+ (-G15)*V17 + (+G15+G16)*V18 + (-G16)*V19 = 0
+ (-G16)*V18 + (+G16+G17)*V19 + (-G17)*V20 = 0
+ (-G18)*V14 + (-G17)*V19 + (+G17+G18)*V20 = 0
+ (-G19)*V14 + (+G19+G20)*V21 = 0
+ (+1)*V5 + (-1)*V6 = -E21
+ (+1)*V8 + (-1)*V9 = -E22
+ (+1)*V15 + (-1)*V16 = -E23

• Potential at nodes	• Currents and voltages sides	
V1=-31.752799	U1=31.752799	I1=0.365521
V2=-43.259396	U2=11.506597	I2=0.365521
V3=-57.299053	U3=14.039657	I3=0.365521
V4=-58.125131	U4=0.826077	I4=0.365521
V5=-335.922113	U5=277.796983	I5=0.108940
V6=864.077887	U6=533.914906	I6=0.108940
V7=330.162981	U7=36.309700	I7=0.108940
V8=293.853281	U8=36.309700	I8=0.108940
V9=2393.853281	U9=164.468886	I9=0.108940
V10=2357.543581	U10=721.801534	I10=0.108940
V11=2193.074695	U11=1.873768	I11=0.108940
V12=1471.273161	U12=43.867956	I12=0.108940
V13=1469.399393	U13=0.726124	I13=0.256581
V14=1425.531437	U14=1.424024	I14=0.256581
V15=-58.851254	U15=0.251449	I15=0.256581
V16=1441.148746	U16=11.463264	I16=0.256581
V17=1439.724722	U17=0.130856	I17=0.256581
V18=1439.473272	U18=2.347715	I18=0.256581
V19=1428.010008	U19=511.729234	I19=0.365521
V20=1427.879152	U20=913.802203	I20=0.365521
V21=913.802203	U21=-1200.000000	I21=0.108940
V22=0	U22=-2100.000000	I22=0.108940
	U23=-1500.000000	I23=0.256581

- **Balance of powers**
  - The power generated: 744.373309
  - The power consumption: 744.373309

- **Nodal analysis results**

V1 = -31.752799
V2 = -43.259396
V3 = -57.299053
V4 = -58.125131
V5 = -335.922113
V6 = 864.077887
V7 = 330.162981
V8 = 293.853281
V9 = 2393.853281
V10 = 2357.543581
V11 = 2193.074695
V12 = 1471.273161
V13 = 1469.399393
V14 = 1425.531437
V15 = -58.851254
V16 = 1441.148746
V17 = 1439.724722
V18 = 1439.473272
V19 = 1428.010008
V20 = 1427.879152
V21 = 913.802203
I21 = 0.108940
I22 = 0.108940
I23 = 0.256581

ANCIA is based on the modified nodal method (MNM) which is, due to its flexibility, the most used and the most efficient method of simulating the electrical circuits on computer. [7, 8] The system of the equations that describes the equivalent circuit as ventilation scheme is a nonlinear algebraic system because the constants of pressure have the expressions:

$$\Delta H_K = R_{aK} Q_K^2 \quad \left(U_K = R_K I_K^2 \right)$$

For every nonlinear resistance, we considered the linear section of the characteristic U-I, so the circuit obtained is a linear resistive circuit.

In the example given, we found that the method used is good for calculating the ventilation systems of different types of electrical machines.

13. THE THERMAL CALCULATION. THE HEAT TRANSFER

Our thermal model is based on the analyses of the characteristic circuit providing almost instant calculation speeds. This allows the user to perform any calculation on time. Alternative numerical methods, for example, the analysis of finite elements [FEA] and calculable dynamic fluids [CFD], normally require some days/weeks to put together a three-dimensional model. The numeric mode may then take some hours (days for a complex problem of transit) to calculate a solution. [13]

The characteristic diagram (the characteristic circuit) is like an electrical scheme:

- The thermal resistance instead the electrical resistance
- Sources of power instead of current sources
- Thermal capacity instead electrical capacity
- Nodal temperatures instead voltages
- Power flow through the resistance instead the current.

All thermal resistances and thermal capacity from the model Motor-CAD are automatically calculated from geometric dimensions and materials properties. The user must be familiar with complex phenomena of heat transfer such as the dimensional analysis of conversion. Motor-CAD automatically selects and solves most appropriate formulation for a given surface and selected cooling time. Motor-CAD configure s efficient, precise and robust mathematical algorithms for forced and natural convection, for cooling liquid, for radiation and conduction. A vast library of lamination correlations and demonstrated turbulent convections is used to give precise models for all internal and external surfaces. The air gap model includes laminar, vortex and turbulent convection.

The calculation in stationary regime

Motor-CAD simplifies the analyses of output dates of the stationary regime by using schematic diagrams, simple and easy to read. [13]

In "Figure 14" we observe the solved main scheme. This scheme is used very much by post processing calculation in stationary regime. The resistances and the power sources indicated in scheme are color coded as machine components shown in editors of cross sections. The vertically oriented resistances represent radial heat transfer and horizontally oriented resistances represent axial heat transfer. The resistances which contain the letter "C" are resistances of convection and resistances which contain the letter "R" are resistances of radiation. Two colors resistances represent interface resistances between two components.

The diagram also shows labels for each of its components (resistances and nodes). With a single click of a button the user can change the scheme to indicate the nodal temperature and the thermal resistance or power values, so will avoid elaborate images. The main scheme is very useful for visualizing the place where there may be restrictions on the heat transfer and what can be done to improve cooling.

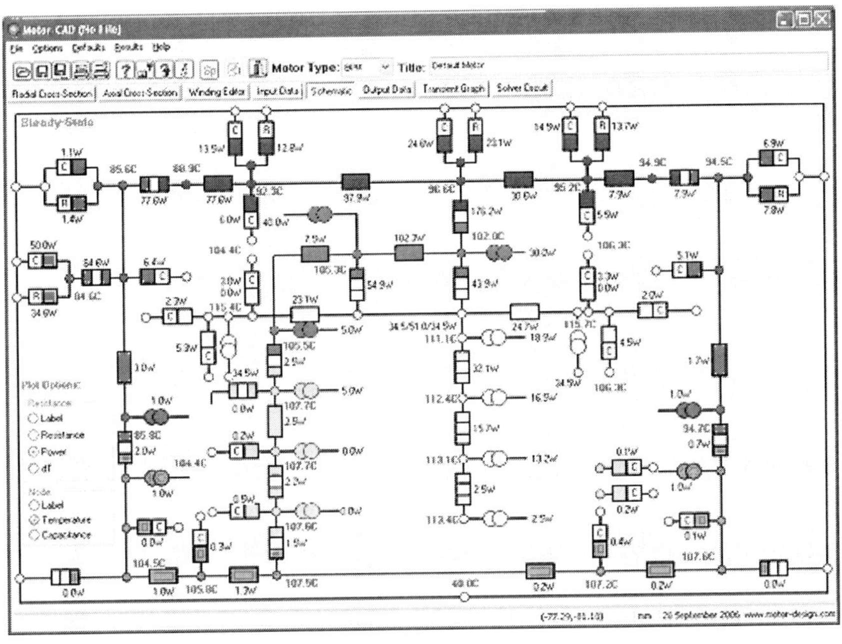

Figure 14. The main scheme (modeling by the characteristic circuit)

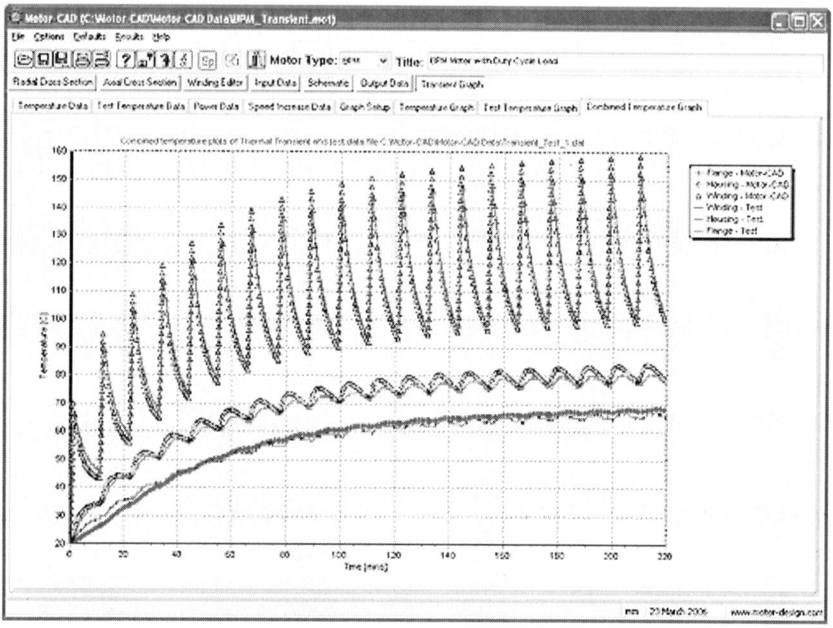

Figure 15. The variation in transient regime

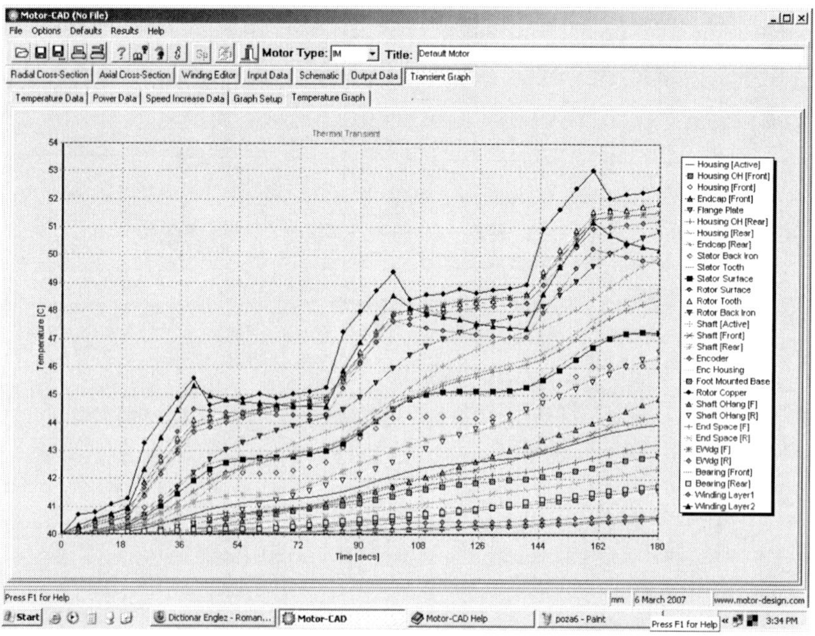

Figure 16. The variation in transient regime on temperature in diverse points in time of the machine

The calculation in transient regime

The Motor-CAD software allows the thermal transition of the motor to be calculated and the outputs to be viewed in graphical format. The opposite graph shows how the test data is important to emphasize on the precision of calculation. Also, Motor-CAD includes facility analyses of the operation cycle. A major advantage of this is that it allows the motor to be directed to its full potential without overheating.

The transient regime can be modeled, so in any case the thermal capacities are automatically added to circuit. It is essential to consider the transient thermal analysis when using operational complex loads if the motor will be used to its full potential. The graph below shows a typical transient regime. In this case we have a repetitive duty cycle (three overloads followed by half loads).

The Motor-CAD software incorporates mathematical efficient and suitable algorithms for the cooling methods. All these, and facility of prescribe miscalculation, if is necessary, provides customization calculations so as to suit the unique construction methods.

14. CONCLUSIONS

Determining or providing the amount of exchanged heat per unit in time in given conditions of temperature is a problem to be taken into account for calculating heating and ventilation for the electrical machines. Were presented the basic sizes of the heat transfer studied in the literature and the distinct forms of heat transfer, the conduction, the radiation and the convection, the basic equations and rules of conduction, radiation and convection. Between the thermal sizes and the electrical sizes we showed a correspondence using the method of thermal resistances and concentrated losses.

By analogy between a network of cooling ventilation and a nonlinear resistive circuits were determined the values of the cooling fluid flows of an electrical machine.

Also by analogy, the temperature drop in the mass or on the surface of an object, corresponding the caloric flow (or loses), was determined by the product between the thermal resistance of the object opposed to the caloric flow in the mass or on surface and loses.

Solving the thermal schemes, were done analogous with solving the electrical schemes, as in electrical circuits of conexions between thermal resistances are series, parallel or mixed.

As we said the exhaust heat from the electric machine is as important as and phenomena related to the proper functioning of the machine. Choosing the most appropriate ventilation is done even by the assignation of electromagnetic stresses.

Were showed the equivalent resistances,andthe aerodynamic schemes used were solved according to the conexions between aerodynamic resistances (series, parallel or mixed).

The discussed example for axial-radial bilateral ventilation in an electrical machineunder protected construction for which the air circulation is symmetrical towards the mid of machine and it is sufficient to calculate ventilation only for one part, which is half of the machine.

After calculating the aerodynamic resistances, were calculated the pressures and air flows using the computer program PACEN (Program of Analysis of Nonlinear Electrical Circuits), and were obtained the system solutions.

Motor-CAD is an example for the analytical analysis of the characteristic circuit, is a program dedicated to electrical motors and generators and has all the indications on limiting design effort.

Motor-CAD has some advantages over CFD:

- The problem of defining time (minutes to hours/days/weeks)
- The calculating rate (instantly to hours/days)
- The postprocesing time (instantly to hours)
- Easy to use.

Motor-CAD allows optimization of the cooling system and identifies vulnerable points of design and/or problems of fabrication; it checks if a provided motor is suitable for a particular application.

The final conclusion for using Motor-CAD is that the program is a quick methoddesigning for motors and electrical generators of thermal point of view, taking into account the fact that the thermal analogous circuit of the electrical machines is as important as designing the electromagnetical, mechanical.

REFERENCES

1. C. Bâlă, 1982Electrical machines- Didactic and Pedagogical Publishing, Bucharest, Romania.

2. Cioc, N. Bichir, N. Cristea, 1981Electrical machines. Design guideline. Vol. I, II, III.-Romanian Writing Publishing, Craiova, Romania.

3. O. Nedelcu, 2010Modeling of heating and ventilation of electrical machines, Bibliotheca Publishing, Targoviste, Romania.

4. L. O. Chua, P. M. Lin, . , 1975Computer-Aided Analysis of Electronic Circuits: Algorithms and Computational Techniques,Prentice Hall, Englewood Cliffs, New Jersey.

5. Leca, I. Prisecaru, 1994Thermophysical and thermodynamical properties.-Technical Publishing, Bucharest, Romania.

6. L. Dumitriu, M. Iordache, (1998, 1998Modern theory of electric circuits- 1-Theoreticalfoundation, Applications, Algorithms and computer programs, All Educational Publishing, Bucharest, Romania.

7. M. Iordache, L. Dumitriu, 2004Computer assisted analysis of nonlinear analog circuits,Politechnica Press publishing, Bucharest, Romania.

8. M. Iordache, L. Dumitriu, 1999PANCIA- Program Analysis of Analog Circuits, User manual, Politechnica Press publishing, Bucharest, Romania.

9. W. J. Mc Calla, 1988Fundamentals of computer-aided circuit simulationKluwer Academic Publishers, Boston.

10. Fl. Hănţilă, (1979, 1979A Method for Solving Nonlinear Resistive Networks, Rev. Roum. Sci. Techn. Électrotechn. et Énerg., 24, 2,2130

11. O. Nedelcu, M. Iordache, D. Enescu, 2006An efficient method for computing of the electric machine ventilation"- The Sixth WESC TORINO, ITALY, July 10-12, 108887380511251

12. O. Nedelcu, D. Enescu, C. I. Sălisteanu, 2009Determination of temperature field distribution and rate of heat transfer in DC machine- WESAS Genova, 978-9-60474-130-4

13. Motor-Cad help

6
CHAPTER

Thermal Interaction Between a Human Body and a Vehicle Cabin

Dragan Ružić and Ferenc Časnji

[1] *University of Novi Sad, Faculty of Technical Sciences, Serbia*

1. INTRODUCTION

The main functions of a vehicle cabin are to provide a comfortable environment for its occupants and to protect them from vibrations, noise and other adverse influences.Since the conditions inside mobile machinery cabins (agricultural tractors and construction equipment) affect the health, performance and comfort of the operator, this research focuses on these types of cabins. Modern agricultural tractors are complex, highly efficient systems, and the aim of the development of these systems is the reduction of the negative impact they have on the environment. Main goals are achieving higher fuel economy, lower emission and less soil compaction caused by tractor wheels. There is also a tendency to improve the ergonomics of operator enclosures. An operator enclosure can be treated as workspace andthe conditions inside atractor cab have significant impact on the performanceof the operator, and in that wayon the total efficiency of the *operator-tractor-environment* system as well. From the operator's point of view, tractor cab ergonomics is a key factor in ensuring his optimum working performance, which could easily become the weakest link in the working process. While, on the one hand, the tractor cab offers mechanical protection and the protection from adverse ambient conditions, on the other hand, even under moderate outside conditions,a closed tractor cab acts like a greenhouse and its interior could become unpleasant, unbearable and even dangerous.

The project presented in the report by Bohm et al. (2002), deals with the thermal effects of glazing in cabs with large glass areas. Using the thermal manikin AIMAN,they studied and evaluated the effects of different kinds of glass and design of the windows, as well as the effects of sun protection and insulation glazing. The results showed that neither in severe winter conditions, norin sunny summer conditions, could acceptable climate be obtained with standardglazing in cabs with large glass areas.

In a comprehensive project which was aimed at the reduction of vehicle auxiliary load, done by National Renewable Energy Laboratory, Golden, USA (Rugh and Farrington, 2008), a variety of research methods was used to research and develop innovative techniques and technologies that reduce the amount of fuel needed for the air-conditioning system, by lowering thermal loads. They concluded that the reflection of the solar radiation incident on the vehicle glass is the most important factor in making significant reductions in thermal loads. The use of solar-reflective glass reduced the average air temperature and the seat temperature. The use of reflective shades and electrochromic switchable glazing are also effective techniques for the reduction of the solar energy that enters the passenger compartment. They also found that solar-reflective coatings on exterior opaque surfaces and vehicle body insulation can reduce a vehicle's interior temperature, but they can do this to a lesser extent than solar-reflective glazing, shades, and parked-car ventilation.

Both projects aimed at the reduction of thermal loads in vehicle cabins, primarily considering the advanced techniques of heat rejection and insulation. The research described in the paper by Currle and Maue (2000) deals with the numerical study of the influence of air flow and vent geometrical parameters (area and position of the vents) on thermal comfort of passenger car occupants. The results showed significant influence of the vent area, not only on velocity levels in the cabin, but also on overall as well as local thermal comfort.

This chapter deals with the characteristics of the agricultural tractor cabs that have direct or indirect effect on thermalergonomics of operator enclosures. For this purpose, thermal processes between hot (summer) environment and a tractor cab, as well as between the cab and anoperator are analysed. The hot conditions are more complex and more demanding in terms of achieving comfortable thermal conditions in a tractor cab. The complexity of thermal interactions between a human, a tractor cab and the environment is shown in Fig. 1.

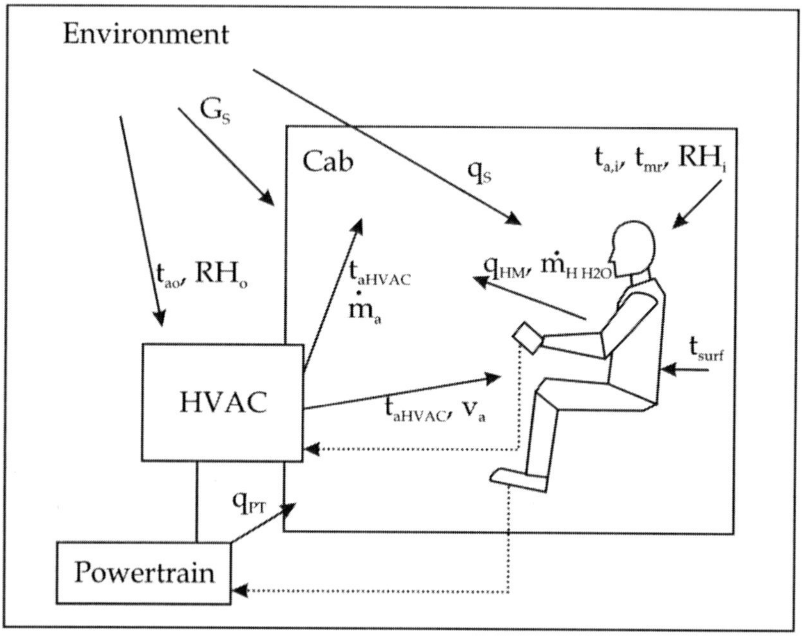

Figure 1. Thermal interaction between a human, a cabin and the environment. Dotted line arrows are control paths

In order to be able to improve thermal ergonomics, it is necessary to have a good knowledge of thermal properties of materials and other design parameters of importance for the cabin thermal processes, as well as the knowledge of the properties of the human body and its thermoregulation system. The main methods for both the improvement of the microclimatic conditions and the reduction of air-conditioning energy consumption in any type of cabin are:

- Passive:
- prevention of cabin heat gain, using principles of heat rejection and insulation;
- prevention of heat transfer to the operator's body by reducing the transferred solar radiation, as well as thermal radiation and heat transfer from the cabin interior surfaces.
- Active: cooling the operator's body by the effective use of airflow from the air-conditioning system.
- The aims of the research are:

- to identify and evaluate the most important influences on heat transfer processes in the human-cabin-environment system,
- to explain the effects of the ventilation (air distribution) system design and settings on the operator's heat loss and thermal sensation in hot conditions.

2. MAIN DESIGN FEATURES OF TRACTOR CABS

An overview of the design features related to the thermal processes and HVAC characteristics wasdone usingrelevant technical documentation and by analysingthedesign of several agricultural tractors (Ružić and Časnji, 2011). Orchard and vineyard tractors (narrow track tractors) were not included in the analysis. All tractors from the sample were 4WD wheeled tractors, with power ranging from 40.5 to 155 kW, their weight ranging from 2750 to 8410 kg and their wheelbase ranging from 2.055 to 3.089 m. The outer lengths of their cabs werebetween 1.40 and 1.77 m, the cab widths were between 1.38 and 1.70 m, and the cab heights were between 1.45 and 1.80 m.

2.1. Materials

The basic materials for tractor cab frames are steel profiles, which primarily must meet the demands for mechanical protection of the operator. Cab frames usually have six pillars, although there are designs with four pillars.

The floor is made of a steel sheet supported by a steel frame. For the purposes of sound and heat insulation, the floor has an interior lining and, in some designs, an outer lining. The powertrain (transmission) of typical tractors is positioned beneath the cab. The powertrain of smaller tractors partly "protrudes" into the cab space, while larger tractors have a flat floor. In the interior,rubber or polymer flooring is used. Some cabs havea foam insulation material between the flooring and the floor. If there is additional heat and sound insulation beneath the floor, compositematerials are used. The typical composite heat shield is made of reflective outer aluminium layer bonded on synthetic fibre core.

Cab roofs are made of polymers, and the cab can be equipped with a roof-window. The components of the ventilation system and the air-conditioning are generally placed in the roof.

Cab glazing in modern tractors takes approximately 60% of the total cab surface area. Glass is generally tempered and tinted.In the sample of analysed tractors, the windshield inclination is in the range of 8 – 20° and side windows inclination is in the range of 7 – 10°.

An overview of thermal properties of materials used in tractor cabs is given in Table 1.

Table 1. Thermal propertiesof tractor cab materials (McAdams, 1969; Incropera andDeWitt, 1981; Siegel & Howell, 1992; ASHRAE, 1997c; Saint-Gobain, 2003)

	thermal conductivity, k (W/mK)	surface emissivity, ε	solar absorptivity, α_s	solar transmissivity normal to surface, τ
tempered single glass, clear	0.8	0.8 – 0.95	0.08	0.84 – 0.90
tempered single glass, green tinted	0.8	0.8 – 0.95	0.45	0.49
metal, painted white	40 – 45	0.85 – 0.96	0.21 – 0.25	-
polymer, white	0.12 – 0.20	0.90 – 0.97	0.23 – 0.49	-
metal, painted black	40 – 45	0.97	0.80	-
rubber, black	0.10 – 0.13	0.86 – 0.92	-	-
insulating foam	0.037 – 0.050	0.95	-	-
composite heat shield	0.040 – 0.060	0.10 – 0.20	-	-

Cabs usually have an air distribution system with air vents (outlets of ventilation duct system) placed on the ceiling. Some of the cabs from the sample also have vents placed on the instrument panel, usually for heating. The air vents on the ceiling can have a symmetrical or an asymmetrical layout, while the vents on the instrument panel always have a symmetrical layout. The vents mostly have a circular cross-section and the operator can change the direction of the air jet.

Vapour compression systems are used for air-conditioning (AC), same as in automotive applications, with R134a as the refrigerant. Installed compressors can take around 4 – 8 kW from the tractor engine, which presents 2.5 – 15% of the rated power in this category of tractors.

3. HEAT EXCHANGE BETWEEN A TRACTOR CAB AND THE ENVIRONMENT

Thermal processes in a tractor cab are in close relation to outside thermal conditions (air temperature, air velocity, intensity and direction of solar radiation). Thermal processes are more or less independent from tractor's working operations, whereasthe noise and vibrations depend on them. Since the decreasein the cab heat gain means the reduction of the thermal load to the operator as well, the analysis was done for adverse environmental conditions that can be encountered over the summer period:

- maximum outdoor air temperature above 30°C,
- total solar irradiation on the outer surfaces up to 1000 W/m^2,
- closed cab so that the operator is protected from noise and air pollution,
- low tractor velocity that does not promote natural tractor cab ventilation.

The sum of heat gain for a closed tractor cab in hot environment, under the steady-state conditions, consists of the following (Fig. 2.):

- heat transfer through the cab envelope due to the temperature difference Q_k,
- heat transfer through the cab roof caused by solar radiation Q_s,
- solar radiation transmission through glazing Q_{GS},
- heat gain from the powertrain Q_{PT},
- sensible (Q_{HM}) and latent heat ($Q_{H\ H2O}$) released by the operator.

Heat load by infiltration is assumed to be zero due to the pressurisation of the tractor cab. In order to maintain the interior temperature constant, the heat removal by the air-conditioning (Q_{AC}) should be equal to the heat gain. According to the requirements for thermal comfort in summer conditions, interior air temperature should be in the range of 23 to 28°C (ASHRAE, 2003).

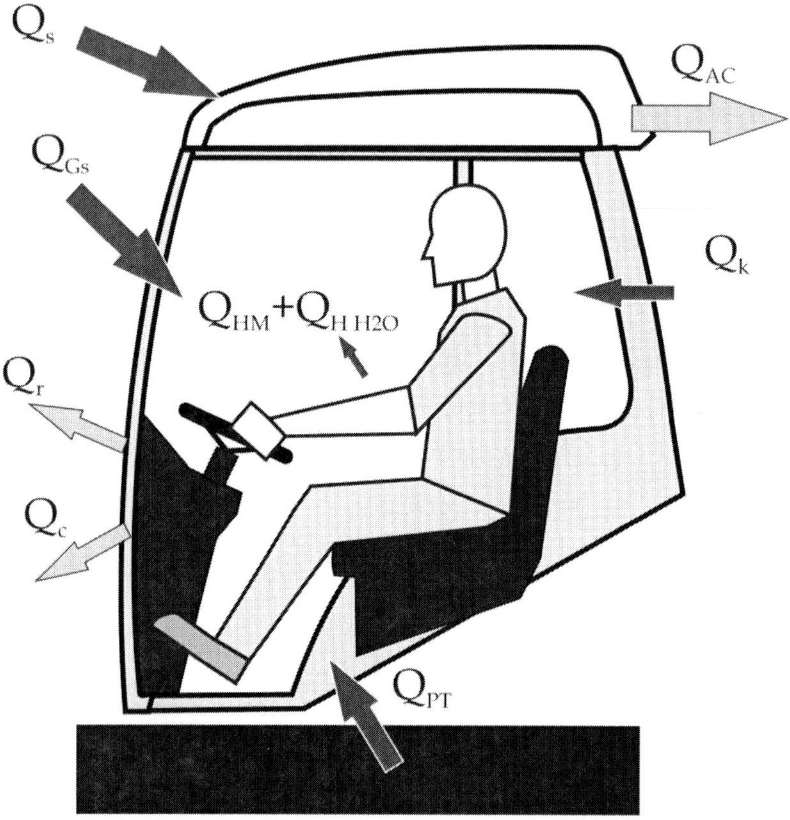

Figure 2. Thermal processes between a tractor cab and hot environment

For an analytical investigation of thermal processes, it was assumed that the heat flow is a one-dimensional steady flow through a plane wall with natural convection on both sides. A problem that comes up inthe analytical determination of heat flux isthe calculation of the convection heat transfer coefficients, which differ according to various sources (Incropera and DeWitt, 1981; ASHRAE, 1997a; Conceicao, 1999; Grossmann, 2010).Radiant heat exchanges among inner surfaces were neglected, because of the complexity of surface shapes and relatively small differences in temperature. The thermal radiation inside the cab is important for the thermal load of the operator, and this will be discussed later.

3.1. Heat Transfer Through Cab Walls Due to Temperature Differences

Assuming that the air temperature in the vicinity of the walls is uniform, the one-dimensional heat load through this part of cab walls is (Fig.3):

$$Q_k = A_k \cdot q_k = A_k \cdot U \cdot (t_{a,o} - t_{a,i}),\ W \quad (1)$$

Parameter U is the total heat transfer coefficient, which combines convections from each side of the wall with conduction. General equation for a wall of area A_j with $s$layers:

$$U_j = \frac{1}{\frac{1}{h_{c,i}} + \sum \frac{\delta_s}{k_s} + \frac{1}{h_{c,o}}},\ W/m^2K \quad (2)$$

For all the surfaces with the heat transfer due to the difference in air temperature, the total value of $U$is:

$$U = \frac{1}{A_k} \sum U_j A_j,\ W/m^2K \quad (3)$$

This value is dependent on the thermal characteristics of the material that walls are made of (thermal conductivity k, W/mK), of the thickness of the layers and the surface area. The surfaces where this equation applies are the surfaces exposed neither to the solar radiation nor to the radiation from the powertrain. For example, these surfaces are windows on the shaded side of the cab.

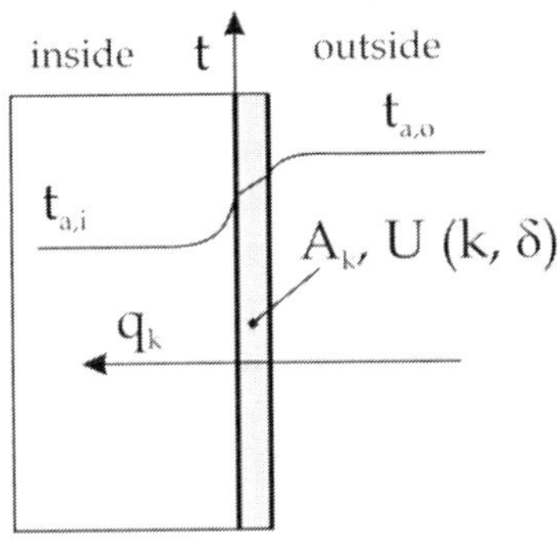

Figure 3. Heat transfer through the cab wall due to air temperature difference.

Based on very few available sources, it is assumed here that the cab of a stationary vehicle has *U*value ofaround 10 W/m²K (Türler, 2003; Karlsson, 2007; Tavast, 2007; Grossmann, 2010). Consequently, the estimated heat flux through the cab wall (q_k) would be around 40W/m²for the difference between the outside and the inside air temperatureof (30 – 26)°C = 4°.

3.2. Effects Of Solar Radiation

Cab glazing is a semitransparent medium, where solar irradiation can be partially reflected, absorbed and transmitted, Fig. 4.

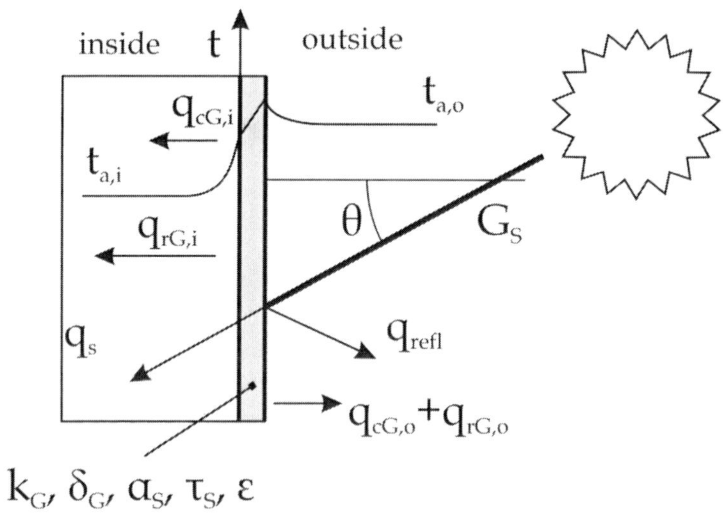

Figure 4. Solar energy transmission through glass

Solar irradiation on a tractor cab surface is variable depending on the position of the sun as well as the orientation of the cab surfaces. An example of maximum ("clear sky") intensity of normal irradiation and its variations in time and the influence of the surface inclination are shown in Fig. 5. The values are given for a vertical surface and for the surface inclined at 20°. The diagram shows that the vertically positioned glass receives around 15% less solar irradiation than the glass inclined at the angle of 20°.

The thermal balance equation for the outer glass surface can be written as:

$$G_s \cdot \alpha_s - \frac{k_G}{\delta_G}\left(t_{G,o} - t_{G,i}\right) - h_{G,o}\left(t_{G,o} - t_{a,o}\right) - \varepsilon \cdot \sigma\left(T_{G,o}^4 - T_{sky}^4\right) = 0 \qquad (4)$$

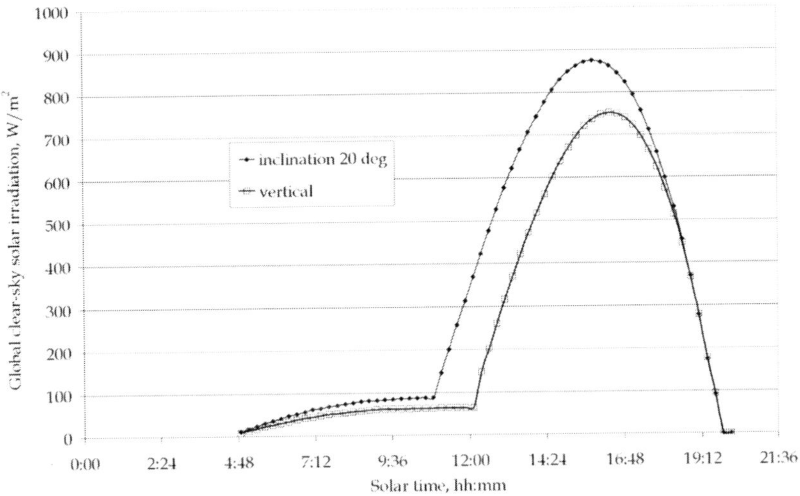

Figure 5. Global solar clear-sky irradiation G_s on a west-faced vertical plane and on a plane inclined at the angle of 20° from vertical, in a central European region on a day in July (http://re.jrc.ec.europa.eu/pvgis/apps3/pvest.php)

The thermal balance on the inner glass surface is:

$$\frac{k_G}{\delta_G}\left(t_{G,o}-t_{G,i}\right)-h_{G,i}\left(t_{G,i}-t_{a,i}\right)-q_{rG,i}=0 \tag{5}$$

The total heat flux through the glass exposed to the solar radiation consists of the transmitted part of the solar radiation, the convection on the inner side and longwave radiation (neglected here due to a small temperature difference between interior surfaces):

$$q_s=G_s\cdot\tau_s+h_c\left(t_{G,i}-t_{a,i}\right),\ W/m^2 \tag{6}$$

Using available values of the thermal properties of glass (Table 1) with the solar irradiation on inclined glass equal to 876 W/m², the estimated total heat flux through the glass will be around 740 W/m² for clear glass, and around 530 W/m² for tinted glass (green, with 75% transmittance of visible light; Saint-Gobain Sekurit, 2003). The air temperature difference was the same as above, i.e. 4°C. The drawback of use of the tinted glass is its solar radiation absorptivity, which causes the rise in the temperatureof glass. In this example, the calculated glass temperatures will be around 30°C and

50°C, for clear and tinted glass respectively. In comparison to solar absorbing glass, a better but also a more expensive solution is the infra-red reflective glass that rejects almost a half of the solar radiation energy with less obstruction of visible light transmission (Bohm, 2002; Türler, 2003; Saint-Gobain, 2003). This kind of glass is used in automotive applications.

The total amount of the heat transmitted through the glass caused by solar radiation is related tothe normal projection of the tractor cab in the direction of radiation.The maximum solar transmissivity of glass is in the region of incident angles that are less than 30°, being the most inconvenient case both in terms of solar radiation transmittance to the cab interior as well as the direct effect on the operator.

The absorbed part of the solar irradiation heats the cab's outer opaque surfaces (the roof, for example) and their temperature rises. The maximum solar irradiation on a horizontal roof surface in the centralEurope region on a summer day may exceed 900 W/m². After some exposure to solar radiation, the thermal balance between heat gain and heat release will be achieved. The heat from the outer surface is released into the environment (mostly to the sky) by longwave radiation, and into the surrounding air by convection. From the inner surface, the heat from the heated wall of the cabin is realeased into the air by convection and it is emitted by longwave radiation.The cab roof is an example of a surface where this mode of heat transfer takes place, Fig. 6.

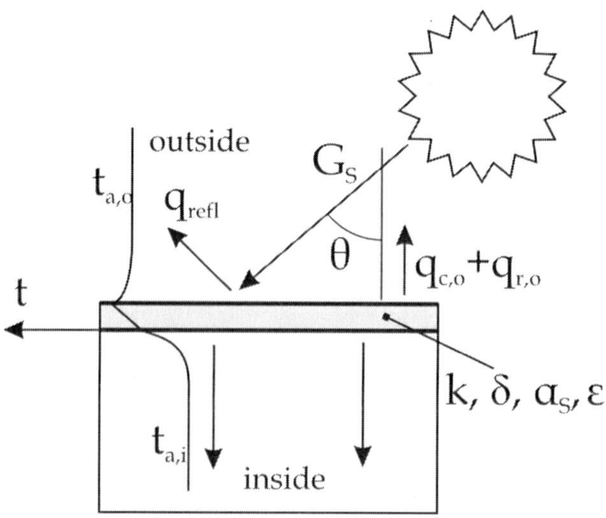

Figure 6. Heat transfer through the cab roof caused by solar radiation.

If there are any components of a ventilation system and an AC system in the roof,the heat transfer from the hot outer roof surface is not direct, but is transferred through the air space.The thermal balance equation for the outer surface of the roof can be written as:

$$G_s \cdot \alpha_s - \varepsilon \cdot \sigma\left(T_{roof,o}^4 - T_{sky}^4\right) - h_{c,o}(t_{roof,o} - t_{a,o}) - \frac{k_{roof}}{\delta_{roof}}(t_{roof,o} - t_{roof,i}) = 0 \quad (7)$$

The thermal balance on the inner roof surface is:

$$\frac{k_{roof}}{\delta_{roof}}\left(t_{roof,o} - t_{roof,i}\right) - h_{roof,i}\left(t_{roof,i} - t_{a,i}\right) - q_{r,roof,i} = 0 \quad (8)$$

Using an example of the heat flux assessment for an opaque roof, treating the roof as a single layer wall, design parameters of the roof will be discussed. The radiation heat exchange between interior surfaces is neglected here, supposing that temperature differences among the surfaces are small. Hence, the equation of heat flux that flows through the roof, caused by solar radiation, is:

$$q_s = \alpha_s \cdot G_s - \varepsilon \cdot \sigma\left(T_{roof,o}^4 - T_{sky}^4\right) - h_{c,o}(t_{roof,o} - t_{a,o}) =$$
$$= \frac{k_{roof}}{\delta_{roof}}(t_{roof,o} - t_{roof,i}) = h_{c,i}(t_{roof,i} - t_{a,i}) \quad (9)$$

For example, the normalized data of heat flux and relative surface temperature, for 800 W/m^2 of solar irradiation on a horizontal surface, are given in Table 2. Four designs are analysed here, the cabs with roofs made of metal or polyester, both with and without a thermal insulating material on the inner side.Solar absorptivity of the outer surface is 0.3and the AC keeps the interior air temperature at 26°C.

As it was in the case with cab glazing, irradiation and solar absorptivity coefficient are dependent on the solar radiation incident angle, θ. The worst case is, of course, at noon, when the radiation is almost normal to the surface.

Table 2. Evaluated relative values of the heat transfer parameters for different roof materials

	Metal	Metal with insulation	Polyester	Polyester with insulation
Heat flux through the roof	100% (~30 W/m²)	70%	90%	60%
Inner surface temperature reduction	0 (surface temperature ~45°C)	6°	2°	7°
Inner surface radiation	100%	94%	99%	94%

Solar absorptivity is also dependent on the colourof the surface.In terms of colour, light colours are preferable, and are widely used for tractor roofs. Under the same conditions, but with a dark coloured roof with solar absorptivity of 0.9, the heat transfer rate increases three times in comparison with the roof that has the solar absorptivity of 0.3. Also, the temperature of the roof interior will be 18 – 30°C higher, while the radiant heat flux emitted by the roof's inner surface would in this case rise more than 50%. Despite the high thermal conductivity, one of the most popular materials for the reflection of thermal radiation is aluminium (α_S = 0.09 – 0.15; Incropera and DeWitt, 1981; Siegel & Howell, 1992), but it is not used for these purposes in tractor cabs.

Obviously, the solar absorptivity and longwave emissivity of the roof material, as well as thermal conductivity,are very importantdesign factors. Therefore, the outer roof surface should have a low solar absorptivity coefficient and high emissivity (small α_S/ε ratio is preferable), with low thermal conductivity at the same time. However, in order to reduce thermal radiation towards the operator's body surface, the inner roof surface should have low thermal emissivity. Unfortunately, this is not the case with the kinds of materials usually used for the undersideof the roof.

3.3. Heat Gain From The Powertrain

Heat transfer through the floor is modelled as the sum of the radiation from hot powertrain surfaces and the natural convection from the surrounding air, as in Fig. 7. It is assumed that theair and the outer surfaces of the powertrain have the temperature of 80°C (based on temperature of the transmission oil; datafrom tractor service manuals). In addition to the radiative properties and the temperature of surfaces, geometry also has some influence on the heat exchange. The geometry is described by the view factor. Because the distance between the powertrain and the floor is small, it is assumed here that the view factor is equal to unity (Incropera and DeWitt, 1981).

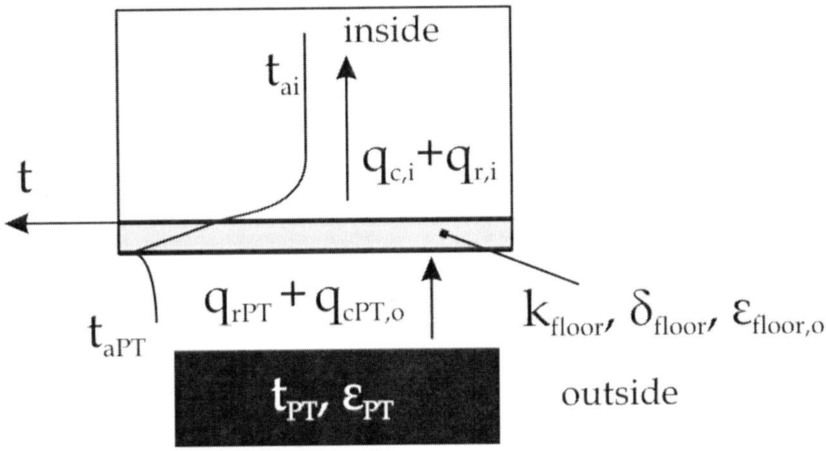

Figure 7. Heat gain from the powertrain

The heat flux that the floor absorbs from the hot powertrain surface and the surrounding air can be written as:

$$q_{PT} = \varepsilon_{PT:floor} \cdot \sigma\left(T_{PT}^4 - T_{floor,o}^4\right) + h_{c,o}\left(t_{a,o} - t_{floor,o}\right),\ W/m^2 \tag{10}$$

The factor $\varepsilon_{PT:floor}$ takes into account different emissivity of involved surfaces (Mc Adams, 1969):

$$\varepsilon_{PT:floor} = \frac{1}{\dfrac{1}{\varepsilon_{PT}} + \dfrac{1}{\varepsilon_{floor,o}} - 1} \tag{11}$$

Under the thermal equilibrium, the same heat flux is transferred through the floor and transmitted to the interior air:

$$q_{PT} = \frac{k_{floor}}{\delta_{floor}}\left(t_{floor,o} - t_{floor,i}\right) = h_{c,i}\left(t_{floor,i} - t_{a,i}\right),\ W/m^2$$

(12)

For a simple floor without a heat shield on the outer side and with rubber flooringonly, the estimated value of the heat flux through the cab floor caused by powertrainis around 110 W/m^2. If there is a heat shield beneath the floor and insulation between the flooring and the metal, the heat gain can be reduced to one third. Consequently, in the first case, the inner temperature of the floor will be above 50°C, while in the second case the temperature will be lower, around 34°C.

3.4. Total Thermal Load Of The Tractor Cab

The total thermal load of the cab under the chosen summer conditions will depend on the size of the cab and on its orientation to the sun. The worst case would be thatthe largest side of the cab is facing the sun and the tractor cab is not equipped with the mentioned means for heat gain reduction. Therefore, for approximately 1/4 of the glazed area exposed to the sun (around 1.5 m^2), the resulting solar heat gainwould be around 1100 W (in tractors with clear glazing). The rest of the cab surface, (approximately more than 7 m^2), would transfer up to 800 W. In total, adding the heat released by the operator, the estimated thermal load of the tractor cab would be around 2 kW. Fig. 8 shows the percentage of different modes of thermal loads and their reduction via the mentioned methods for heat rejection and insulation.The chart in Fig. 8shows that glazing has the most important role in thermal processes that influence the microclimate conditions inside the tractor cab.

4. THERMAL PROCESSES BETWEEN AN OPERATOR AND A TRACTOR CAB INTERIOR

The conventional approach for the analysis of thermal processes is based on the balance between the sensible and latent thermal load of the cabin and the heat removal by supplied air. Therefore, quantity and conditions of the air supplied by the AC influences the overall thermal state in the cabin, as shown in Fig. 9. This approach, although technically correct, is not suitable for the assessment and prediction of the operator's thermal sensation. Due to the complex nature of the thermal environment in a

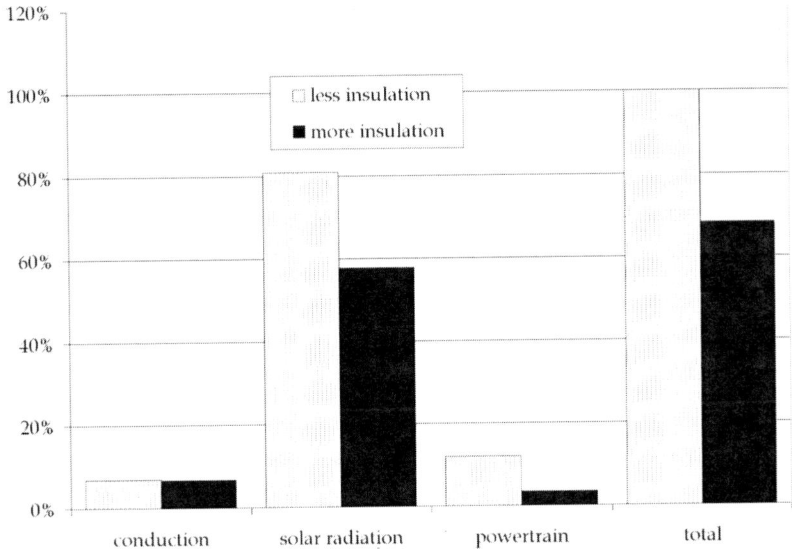

Figure 8. Percentage of heat fluxes of different modes of heat transfer to the tractor cab and the effects of application of the heat rejection and insulation

tractor cab, it is not possible to accurately describe the thermal conditions inside a tractor cab by a single value, even when the value combines the microclimate factors (for example PMV index; Fanger, 1970). Because of this, other assessment parameters have been developed. One of them is the equivalent temperature (t_{eq}), defined as the temperature of a homogenous space, with mean radiant temperature equal to air temperature and with zero air velocity, in which a person exchanges the same heat loss by convection and radiation as in the actual conditions. The equivalent temperature can be evaluated for the whole body or for individual body parts.

In the conventional approach, an operator is treated only as a source of certain amount of sensible and latent heat. The human-based design takes into account the operator's shape, volume, and his thermal sensitivity.

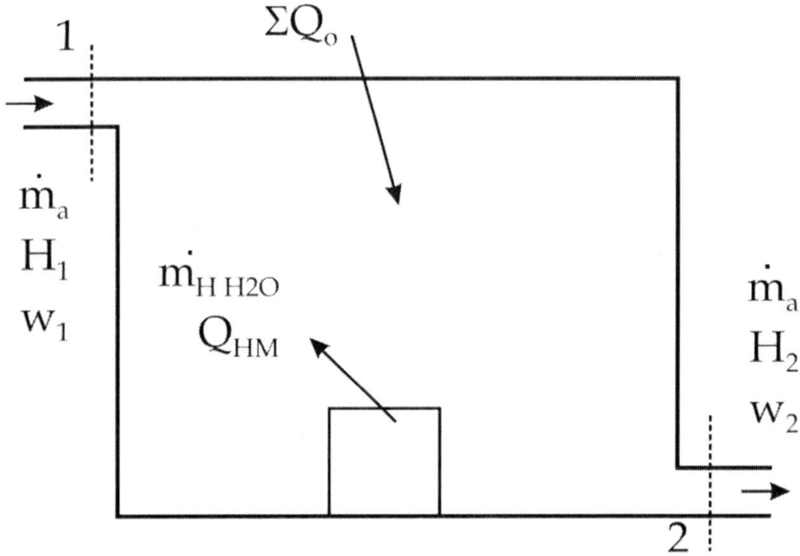

Figure 9. Schematic of the conventional approach to the analysis of air-conditioned space, considering only the air conditions at the inlet (1) and the outlet (2) and the incident rates of energy and moisture gains

4.1. Human Body Thermoregulation And Thermal Sensation

Microclimate conditions and hence human thermal sensation as well, are dependent on air temperature, air velocity, relative humidity and mean radiant temperature. However, individual differences regarding physiological and psychological response, clothing insulation, activity and preferencesin terms of air temperature and air movement also have a strong impact on thermal sensation.

In an uncomfortable hot ambient, owing to the thermoregulation system, the human body sets off the process of vasodilatation and sweating, trying to prevent the rise of internal body temperature. Since these processes can cool down the body only to a certain degree, in order to prevent further rise in body temperature and to avoid the risk of hyperthermia, it is necessary to make the space comfortable or to enhance the process of releasing the heat from the body. Both of these methods are used in vehicles, the latter being more suitable for transient (cool-down) conditions.

The heat loss from the body surface in warm conditionsrelies on convective heat loss and the heat loss by sweat evaporation from the skin surface (Fanger, 1970; Parsons, 2003).The equation for the dry heat flux by convectioncan be written as:

$$C = f_{cl} \cdot h_c \cdot (t_{cl} - t_a),\ W/m^2 \tag{13}$$

Factor f_{cl} is the clothing area factor, which describes the ratio between the area of the clothed body and the area of the nude body. The temperature of theclothing surface (t_{cl}) will be equal to the skin temperature (t_{sk}) on body surfaces without clothing: face, forearms, hands and neck.

According to the equation (13), the higher the temperature difference between the air and the body surface, the higher the heat loss from the body. In addition, the coefficient of heat transfer by convection (h_c) is dependent on air velocity and the shape of abody part. The coefficient increases with the increase of air velocity (Fanger, 1970; De Dear et al., 1997; Parsons, 2003). Therefore, the ventilation system is able to influence this part of the heat transfer directly, by changing the airflow rate through the ventilation outlets (air vents). The general form of the equation for forced and natural convective heat transfer coefficient for the whole body is (Fanger, 1970; De Dear et al., 1997):

$$h_c = B_f \cdot v_a^n,\ \text{for forced convection, and} \tag{14}$$

$$h_c = B_n \cdot v_a^m,\ \text{for natural convection.} \tag{15}$$

The exponents *n* and *m* and the constants B_f and B_n are mostly dependent on body postureas found in various literary sources. The overview of the equations for the calculations of the heat transfer coefficient can be found in Fanger (1970), ASHRAE (1997b), De Dear et al. (1997) and Parsons (2003).

Evaporative heat loss from the skin is dependent on air humidity, skin wetness and evaporative heat transfer coefficient (h_e). The evaporative heat transfer coefficient is calculated using the Lewis relation (Parsons, 2003):

$$h_e = LR \cdot h_c,\ W/m^2kPa \tag{16}$$

Lewis ratio (LR = 16.5 K/kPa) makes the evaporative heat transfer coefficientdependent on air velocity as well. In addition, air-conditioning lowers the relative humidity of the interior air and consequently improves the latent heat loss by the evaporation of sweat.

The body's overall thermal sensation is affected by the local thermal state of individual body parts, at the same level of cooling, due to different physiological properties (different sensibility for warm versus cool sensations, different sweating rate etc.; Zhang, 2003):

- Back, chest, and pelvis strongly influence overall thermal sensation, which closely follows the local sensation of these parts during local cooling;
- The head region, arms and legs have an intermediate influence on the body's overall thermal sensation;
- Hands and feet have much less impact on the overall sensation.

4.2. Local Air Velocities Around A Human Body

The working principle of the vehicle air-conditioning and the ventilation system is based on drivingthe conditioned air through adjustable air vents in the vehicle cabin. This airflow causes changes in local and overall microclimate conditions, consequently also changing the heat loss from the operator's body. This airflow is characterized by spatially distributed local air velocities and air temperatures.

Local velocity ofthe airflow should have the ability to penetrate the operator's free convection flow. For example, Melikov(2004) recommends local air velocities around 0.3 m/s as the minimum values. ASHRAE 55-2009 standard suggests elevated air velocities up to 1.2 m/s, but only if the air speed is under the control of an exposed person and in the temperature range of 28 - 31°C. These values apply to a lightly clothed sedentary person (0.5 - 0.7 Clo), with the metabolic rate in the range between 1.0 and 1.3 Met. In the case of higher difference between the temperatures of ambient air and (colder) local airflow, the operator will prefer lower air velocities.On the other hand, higher metabolic activity, up to 3.2 Met for the operation of mobile machinery, could allow higher air velocities (ASHRAE, 2003;Melikov, 2004; Arens et al., 2009).

Other very important parameters for thermal sensation and the sensation of air qualityare characteristics of the air in the operator's breathing zone. The breathing zone is a semi-spherical space around the mouth and the nose. In terms of perceived air quality, still air in the breathing zone is not acceptable in a warm ambient, even when overall comfort is achieved. The temperature of the localized air in the breathing zone should be equal or 3 - 4°C lower than the interior air temperature. In order to achieve the mentioned requirements, it is recommended to place air vents at distances from 0.4 to 0.6 m from the operator's breathing zone(Melikov, 2004). This should be done in agreement with the geometry of the interior space of the vehicle cabin.

4.3. Effects. Of Thermal Radiation And Transmitted Solar Radiation

When a tractor cab is exposed to the sun, the operator's body receives the heat partly by solar radiation transmitted through the glass and partly by longwave thermal radiation from the surrounding surfaces. The amount of solar radiation energy that will be absorbed by the body will depend on the effective projected area and on the solar absorptivity of the body surface. The largest effective projected radiation area A_{eff} of a person in the sitting position is withthe azimuth and altitude angles of 30° and 15° respectively, and the total surface is equal to (Fanger, 1970; Parsons, 2003):

$$A_{eff} = 0.72 \cdot 0.33 A_{Du} = 0.238 A_{Du}, \ m^2 \quad (17)$$

The total surface of the human body (A_{Du}) can be calculated from the body weight and height (after Dubois; Fanger, 1970; Parsons, 2003). The factor 0.72 in equation (17) is the effective radiation area factor, and the factor 0.33 is the projected area factor (Fanger, 1970; Parsons, 2003). Solar absorptivity of the human skin is around 0.62, while solar absorptivity for clothing depends on the colour (Incropera and DeWitt, 1981; Parsons, 2003). Tractor cab design and the operator's position on the seat do not offer good protection from solar radiation, although large cabs create better shading for the operator.As it can be seen in Fig. 10, the head and the chest are exposed and protected only by solar properties of glass, unless some kind of solar shading devices are used.Shading devices can be placed on the inner side, or as a better solution, on the outer side of the cab. In both cases, the shading devices must not restrict the operator's normal field of vision in the working area of the tractor. For that reason, sun visors or curtains must be easily adjustable. In addition to shading the operator, the outer shading devices prevent exposure of glass and other surfaces to solar radiation. A common solution for these purposes is the use of cab roof overhangs, since aerodynamics is not an issue in agricultural tractors.

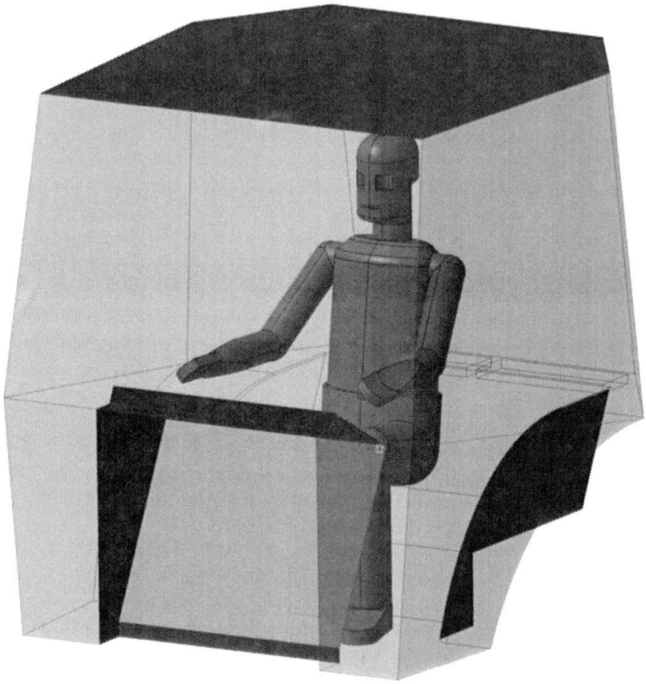

Figure 10. Projection of the operator inside the tractor cab from the direction of the azimuth and altitude angles of 30° and 15°, respectively

When surrounded by surfaces with higher temperatures than that of their skin and clothing, a person receives the heat by thermal radiation. Thermal radiation flux from hot surrounding surfaces depends on the wall surface temperature and the emissivityof the wall. The wall temperature and its emissivity are expressed by mean radiant temperature (t_{mr}) and linear radiative heat transfer coefficient (h_r):

$$q_r = h_r \cdot f_{cl} \cdot \left(t_{mr} - t_{cl}\right),\ W/m^2 \quad (18)$$

$$h_r = 2.88 \cdot \varepsilon \cdot \sigma \left(273.2 + \frac{t_{cl} + t_{mr}}{2}\right)^3,\ W/m^2K \quad (19)$$

The emissivity of non-metallic surfaces is generally high, 0.9 and higher, just like the emissivity of the human skin (0.95;Incropera and DeWitt, 1981;

ASHRAE 1997c) and clothing (0.77 for cotton; Siegel & Howell, 1992). For example, under the thermal conditions mentioned above, in the case of a single-layer metal sheet roof,the thermal radiation load to the operator's head would be around 70 W/m^2. In the case of having the polyester roof with internal thermal insulation, the thermal load would be less than 30 W/m^2.

The AC system is not able to directly change mean radiant temperature, except by decreasing the temperature of the inner surfaces (although very slowly).

5. DESIGN OF THE AIR DISTRIBUTION SYSTEM

The ventilation system drives and distributes the air inside the tractor cab using the air distribution system. The air distribution system consists of a blower, air ducts and several air vents usually positioned on the cab ceiling. The tasks of the ventilation system are:

- to remove excess heat from the operator's body,
- to supply the breathing zone with fresh air,
- to reduce the temperature inside the cab, and
- to pressurize the cab.

The air supply rate of tractor cab ventilation systems are generally in the range of several hundreds of cubic meters of air per hour, and are sufficient for up to 10 kW of cooling power.

According to the air jet theory, the air velocity along the airflow centreline is in direct proportion with the outlet velocity and in inverse proportion with the distance from the air vent. Consequently, the air jet direction, the distance from the body and the velocity profile on the outlet of the air vent are the main factors that influence the characteristics of the airflow over the body surface. These factors are determined by the interior geometry of the tractor cab, the operator's position in relation to the air vents and by the design of the ventilation ducts and vents (cross-section, number of vents), as well.

Control of the microenvironment is necessary due to individual differences among various operators, and airflow characteristics must be suited to individual human thermal sensation. For example, a powerful air-conditioner that easily decreases the temperature inside the tractor cab could produce an unpleasant stream of air, a draught. On the other hand, despite sufficient cooling power of the AC system in some cases, there is the possibility that a part of the operator's body would not be cooled satisfactorily, especially if exposed to solar radiation.

5.1.Evaluation. Of The Air Distribution System's Efficiency And Thermal Sensation

The evaluation of thermal sensation in a tractor cab can be done by using human subjects, by directly measuringthe microclimate physical quantities at discrete points in the cab or by using special human-shaped sensors. Complex human-shaped measuring instruments, the so-called thermal manikins, are the most suitable tool for a reliable, repetitive and objective evaluation of thermal conditions in a non-homogenous environment. Numerical methods for the research of vehicle cabin microclimate are also used. When CFD (Computational Fluid Dynamic) methods for the modelling of the *human-cabin-environment* system are to be used, it is necessary to have a model of the human body that is geometrically and thermally appropriate, the so-called computer simulated person (CSP). Fig. 11.

The analysis of different tractor cab ventilation layouts was described in Ružić et al. (2011) and Ružić (2012). In this research, the heat losses from body parts were observed on a virtual thermal manikin using the CFD simulation. All the simulations were performed without solar radiation. The airflow through the ventilation system was approximately 380 m^3/h, with the air temperature of 20°C. In one group of simulations, the conditions were isothermal (Ružić et al., 2011), while in the other (Ružić, 2012) the initial conditions were the same as the hot ambient with the cab temperature of 30°C.

The results showed that the heat loss from the operator's body can vary significantly depending on the position and the direction of the vents, but all the simulations were done with the same cooling power. Consequently, the operator's thermal sensation will vary accordingly. As it can be seen in Fig. 12, under the same tractor cab thermal load and the same heat removal, but with two different air distribution designs, in the example *A* the total heat loss from the operator is 42% lower than in the example *B*. This indicates that human body characteristics and air distribution systems are two very important factors as well.

In terms of heat loss for the whole body and for thermally most sensitive body segments, the statistical analysisof results for various combinations of vents positioning and airflow direction shows that airflow direction is a more significantparameter than positioning of the vents. Therefore, it is important that the operator be able to change the characteristics of air jet, that is, the air flow direction and the velocity.

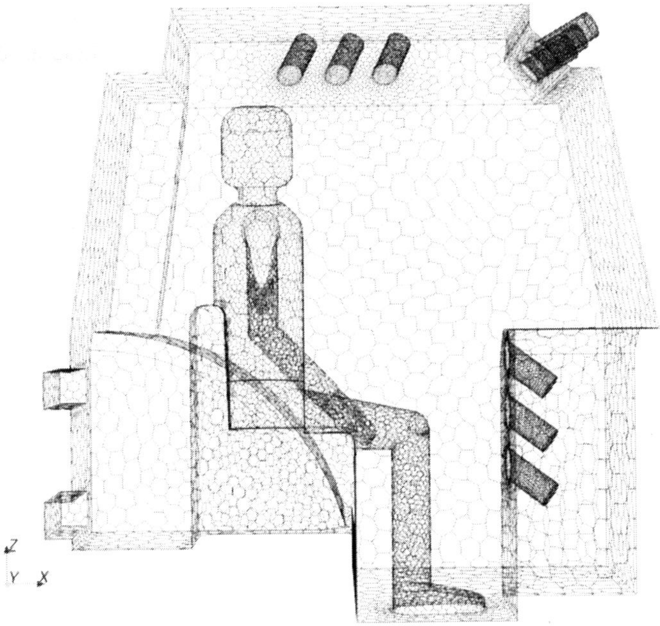

Figure 11. An example of a model of anagricultural tractor cab with a virtual thermal manikin,divided into the volume meshfor the CFD simulation

In the non-isothermal cases (Ružić, 2012), the assessment criterion was the deviation from neutral values of equivalent temperatures. For example, under these conditions, the lowest deviation of the equivalent temperature for the whole body was achieved with the vents placed in the front part of the ceiling, directed at the operator's chest. The highest rate of heat loss from the operator's body was achieved in the layout with the vents placed in the instrument panel, directed at the operator's pelvis (H-point). This mode of operation would be of interest for cooling down of the operator under the conditions of high thermal load, for example caused by solar radiation. The difference between the best and the worst cases regarding equivalent temperature for the whole body was 10 degrees. For individual body segments, the difference was considerably larger. Consequently, if the operator sets the air distribution system in a wrong way, unwanted thermal conditions can be produced. For that reason, the interface for the AC and the ventilation system control must be designed in such a way that it is easy to use.

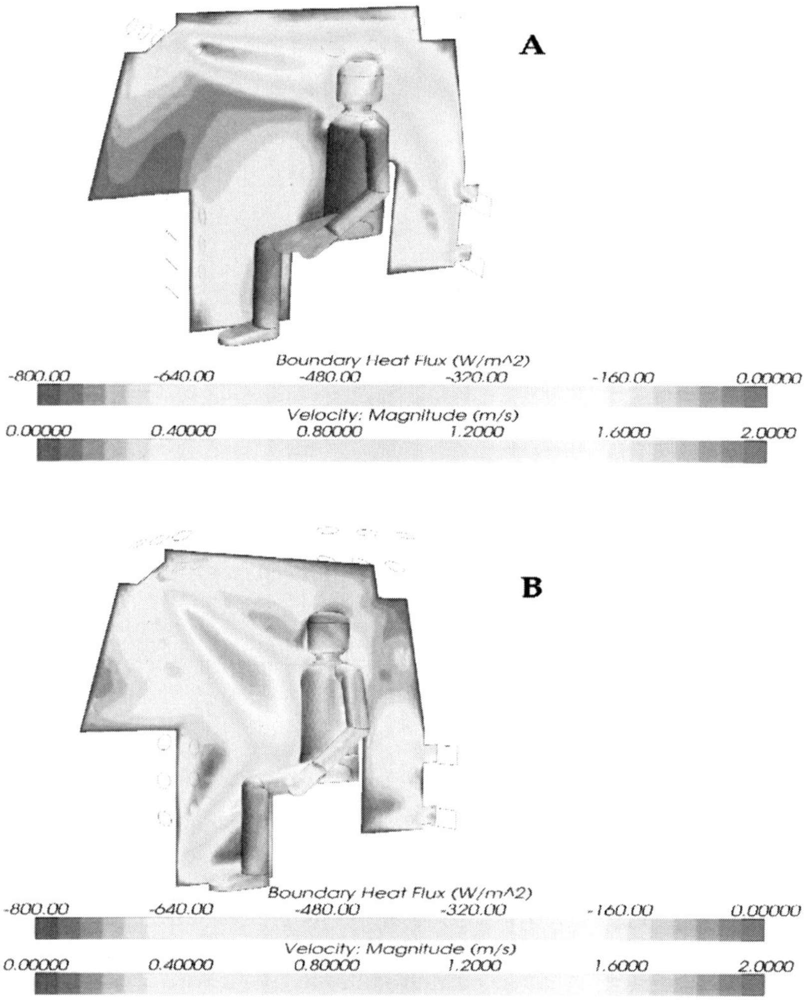

Figure 12. Comparison of air velocity distribution and heat loss from different parts of the operator's body using the CFD technique with virtual thermal manikin (Ružić, 2012)

More information about the significance of certain factors could be obtained by expanding the range of factor variations. However, there is almost an unlimited number of different settings of just one air distribution design, as well as a variety of ambient conditions. Fortunately, the cabs of modern middle-range agricultural tractors are all similar in design and shape, and most of these cabs have the AC air distribution system with the vents on the ceiling. According to the results of those simulations, it is possible to improve and optimize thermal conditions in tractor cabs by implementing conventional air distribution systems.

6. CONCLUSIONS

Although the cabs of modern tractors have many common features, the results show that there are significant differences regarding thermal loads, caused by variations in tractor cab designs. Conclusions can be summarized as follows:

- The highest heat flux that enters the cab is caused by solar radiation through the glass (which is several times higher than the heattransferred by other modes). Paying attention to solar characteristics of glass is a direct way to reduce the operator's thermal load.
- Further improvements in the reduction of the heat load are the use of roof overhangs,adjustable sun visors and less inclined glass. A thermally reflective heat shield placed beneath the cab floor also improves thermal conditions. However, some of the existing technologies for thermal load reduction are not used to the full extent, especially in low-class tractors.
- The ventilation system and the AC system should be designed in such a way that the optimum heat loss from individual body parts can be obtained in most conditions. The direction of the air jet is one of the most crucial factors regarding heat losses from the body.Due to a wide range of boundary conditions and different individual preferences regarding thermal conditions, the air distribution system must be adequately adjustable and easy to use.
- For the analysis and optimization of the air conditioning system, experimentally verified numerical methods (CFD) with computer simulated person or virtual thermal manikin are an inevitable tool.

ACKNOWLEDGEMENT

This research was done as a part of the project TR31046 - "Improvement of the Quality of Tractors and Mobile Systems with the Aim of Increasing Competitiveness and Preserving Soil and Environment", which was supported by the Serbian Ministry of Science and Technological Development.

REFERENCES

1. E. Arens, S. Turner, H. Zhang, G. Paliaga, 2009A Standard for Elevated Air Speed in Neutral and Warm Environments, ASHRAE Journal, 5125May 2009), 8180001-2491
2. ASABE/ISO1997142692Tractors and Self-propelled Machines for Agriculture and Forestry- Operator Enclosure Environment- Part 2: Heating, Ventilation and Air-conditioning Test Method and Performance
3. ASHRAE.1997aFundamentals Handbook, Chapter 3: Heat Transfer, American Society of Heating, Refrigerating and Air Conditioning Engineers, Atlanta, USA
4. ASHRAE.1997bFundamentals Handbook, Chapter 8: Thermal Comfort, American Society of Heating, Refrigerating and Air Conditioning Engineers, Atlanta, USA
5. ASHRAE.1997cFundamentals Handbook, Chapter 36: Physical Properties of Materials, American Society of Heating, Refrigerating and Air Conditioning Engineers, Atlanta, USA
6. ASHRAE.1999aBook of Applications, Chapter 8: Surface Transportation, American Society of Heating, Refrigerating and Air Conditioning Engineers, Atlanta, USA
7. ASHRAE.1999bBook of Applications, Chapter 32: Solar Energy Use, American Society of Heating, Refrigerating and Air Conditioning Engineers, Atlanta, USA
8. ASHRAE.2003Standard 55P: Thermal Environmental Conditions for Human Occupancy, Third Public Review, American Society of Heating, Refrigerating and Air Conditioning Engineers, Atlanta, USA
9. M. Bohm, I. Holmer, H. Nilsson, O. Noren, 2002Thermal effects of glazing in driver's cabs, JTI-rapport 305, JTI- Institutet for jordbruks, 1401-4963 Uppsala, Sweden
10. E. Conceicao, M. Silva, J. Andre, D. Viegas, 1998A Computational Model to Simulate the Thermal Behaviour of the Passengers Compartment of Vehicle, SAE Paper 1999-01-0778, Society of Automotive Engineers, USA
11. J. Currle, J. Maue, 2000Numerical Study of the Influence of Air Vent Area and Air Mass Flux on the Thermal Comfort of Car Occupants, SAE Paper 2000-01-0980, Society of Automotive Engineers, USA

12. R. De Dear, E. Arens, Z. Hui, M. Oguro, 1997Convective and Radiative Heat Transfer Coefficients for Individual Human Body Segments.International Journal of Biometeorology404May 1997), 1411561432-1254

13. P. O. Fanger, (1970). Thermal Comfort, McGraw-Hill, 8-75710-341-0 Denmark

14. H. Grossmann, 2010Pkw-klimatisierung, Springer-Verlag, 978-3-64205-494-5 Berlin, Germany

15. F. P. Incropera, D. P. De Witt, 1981Fundamentals of Heat and Mass TransferJohn Wiley & Sons, 978-0-47142-711-7New York, USA

16. Karlsson, Anika2007Simulation of Cooling Demand in a Truck Compartment, Master thesis, 1650-8300Uppsala Universitet, Sweden

17. W. Mc Adams, 1969Heat Transmission, (in Serbian) Građevinska knjiga, Beograd, Serbia

18. A. Melikov, 2004Personalized VentilationIndoor Air14Suppl. 7), (December 2004), 1571670905-6947

19. K. Parsons, 2003Human Thermal Environments: The Effects of Hot, Moderate and Cold Environments on Human Health, Comfort and Performance, 2nd ed. Taylor & Francis, 0-41523-792-0UK

20. J. Rugh, R. Farrington, 2008Vehicle Ancillary Load Reduction Project Close-Out Report, Technical Report NREL/TP-54042454 , January 2008, National Renewable Energy Laboratory, Golden, USA

21. D. Ružić, F. &Časnji, 2011Agricultural Tractor Cab Characteristics Relevant for Microclimatic Conditions. Journal of Applied Engineering Science 29June 2011), 3233301451-4117

22. D. Ružić, F. . Časnji, A. Poznić, 2011Efficiency Assessment of Different Cab Air Distribution System Layouts, Proceedings of 15th Symposium on Thermal Science and Engineering of Serbia, 819828978-8-66055-018-9Sokobanja, Serbia, October 18- 21, 2012

23. D. Ružić, 2012Analysis of Airflow Direction on Heat Loss From Operator's Body in an Agricultural Tractor Cab, Proceedings of 40th International Symposium on agricultural engineering "Actual Tasks on Agricultural Engineering", 16116913332651Croatia, February 21- 24, 2012

24. Saint-Gobain Sekurit.2003Glazing Manual, available at http://www.sekurit.com,

25. R. Siegel, J. R. Howell, 1992Thermal Radiation Heat Transferrd Edition, Hemisphere Publishing Corporation, 0-89116-271-2USA

26. D. N. Sorensen, L. K. Voigt, 2003Modelling Flow and Heat Transfer Around a Seated Human Body by Computational Fluid DynamicsBuilding and Environment386June 2003), 7537620360-1323

27. J. Tavast, 2007Solar Control Glazing for Trucks, Master thesis, 1650-8300Uppsala Universitet, Sweden

28. D. Türler, D. Hopkins, H. Goudey, 2003Reducing Vehicle Auxiliary Loads Using Advanced Thermal Insulation and Window Technologies, SAE Paper 03HX-36, Society of Automotive Engineers, USA

29. H. Zhang, 2003Human Thermal Sensation and Comfort in Transient and Non-Uniform Thermal EnvironmentsPhD thesis, University of California, Berkeley, USA

7
CHAPTER

A Numerical and Experimental Study of Natural Convective Heat Transfer From an Inclined Isothermal Square Cylinder with an Exposed Top Surface

Abdulrahim Kalendar[1], Patrick H. Oosthuizen[2]

[1]*Department of Mechanical Power and Refrigeration, College of Technological Studies- AAET, Shuwaikh, Kuwait*

[2]*Department of Mechanical and Materials Engineering, Queen's University, Kingston, ON K7L 3N6, Canada*

ABSTRACT

Natural convective heat transfer from an isothermal inclined cylinder with a square cross-section which have an exposed top surface and is, in general, inclined at an angle to the vertical has been numerically and experimentally studied. The cylinder is mounted on a flat adiabatic base plate, the cylinder being normal to the base plate. The numerical solution has been obtained by solving the dimensionless governing equations subject to the boundary conditions using the commercial cfd solver, FLUENT. The flow has been assumed to be symmetrical about the vertical center-plane through the cylinder. Results have only been obtained for Prandtl number of 0.7. Values of inclination angle between 0° and 180° and a wide range of Rayleigh number and the dimensionless cylinder width, $W = w/h$, have been considered. The effects of Dimensionless widths, Rayleigh numbers, and inclination angles on the mean Nusselt number for the entire cylinder and for the mean Nusselt numbers for the various surfaces that make up the cylinder have been examined. Empirical equations for the heat transfer rates from the entire cylinder have been derived.

1 INTRODUCTION

Natural convective heat transfer from an isothermal inclined cylinder with a square cross-section which have an exposed top surface and is, in general, inclined at an angle to the vertical has been numerically and experimentally studied. Some electrical and electronic component cooling problems can be approximately modeled as involving natural convective heat transfer from an isothermal inclined cylinder with a square cross-section mounted on a flat adiabatic base plate. The cylinder has an exposed top surface which is also isothermal and has the same temperature as that of the vertical heated side surfaces. This flow situation, which is considered here, is shown in Fig. 1. The heat transfer from the surface of the cylinder at various angles of inclination between vertically upwards and vertically downwards, as in Fig. 2, has been numerically and experimentally investigated. One of the main aims of the present work was to determine how the cross-sectional size-to-height ratio of the square cylinder, i.e., w/h, influences the mean heat transfer rate from the cylinder at various angles of inclination.

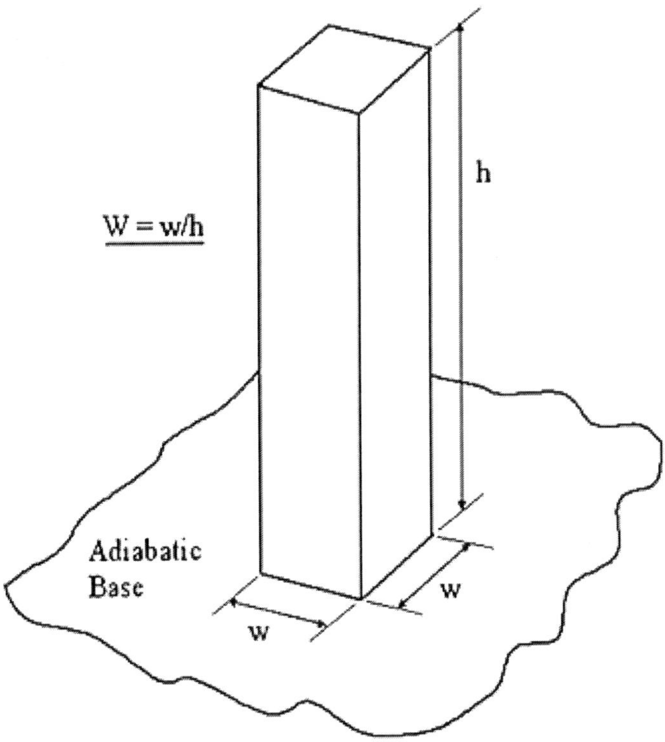

Figure 1. Flow situation considered

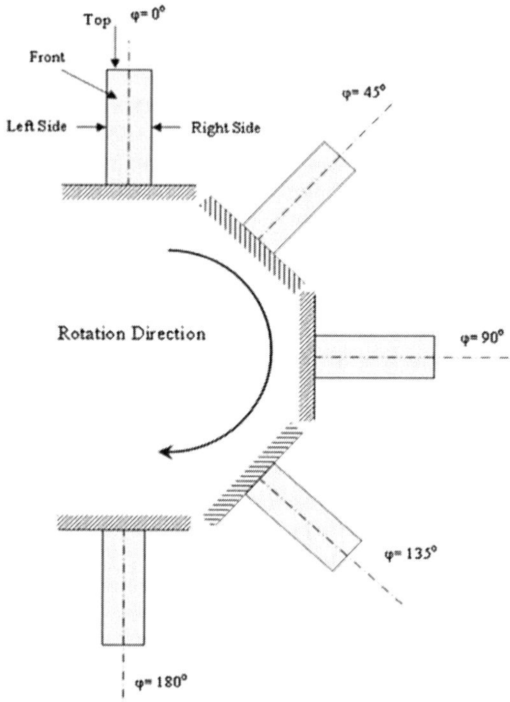

Figure 2. Definition of inclination angle and faces

Natural convection from vertical slender circular cylinders has been studied for many years. Reviews of early work were provided by Ede [1], Burmeister [2], and Jaluria [3]. Cebeci [4] gives numerical results for Prandtl numbers from 0.01 to 100, extending earlier work that had been for Pr = l and 0.72. This work was further extended by Lee et al. [5]. Integral method results were obtained by LeFevre and Ede [6]. The effect of Prandtl number was further studied by Crane [7]. There have been several more recent studies that have further extended this earlier work typical of these being that of Kimura et al. [8] and Oosthuizen [9, 10]. Typical of the experimental studies of natural convective heat transfer from vertical circular cylinders are those of Jarall and Campo [11], Welling et al. [12] and Fukusawa and Iguchi [13]. Cylinders with exposed top inclined at an angle to the vertical have received attention by Oosthuizen [14, 15].

Some studies of natural convective heat transfer from inclined cylinders have been undertaken. In very early study such as Koch [16] and Farber and Rennat [17]. Few studies however have been concerned with short inclined cylinders such as Oosthuizen [18], Eckert and Jackson [19], Al-

Arabi and Salman [20], Al-Arabi and Khamis [21], Chand and Vir [22], Oosthuizen and Mansingh [23] and Oosthuizen and Paul [24].

Some studies of natural convective heat transfer from inclined square cylinders have been undertaken. An experimental study of natural convective heat transfer from an isothermal vertical short and slender square cylinder to air, where the top and bottom surfaces were insulated, were obtained using a lumped capacitance method by Popiel and Wojtkowiak [25]. Even with their widest square cylinder, the mean Nusselt numbers were higher than given by the correlation equation for the wide flat plate, this being especially true at low Rayleigh numbers. Experiments were carried out by Al-Arabi and Sarhan [26] to measure the average natural convection heat transfer rate from the outside surface of isothermal square cylinders of different length and side-lengths at different angle of inclination between 0° and 90° to the vertical in both laminar and turbulent regions. The heat transfer rates from the side surfaces of the square cylinder where the top and bottom ends are insulated were measured. The heat transfer rate was found to depend on side-length and the inclination angle. They found that due to the presence of edge effects, the heat transfer coefficient always exceeded that which would occur under the same conditions with a vertical plate. They also suggested general correlation equations. Natural convective heat transfer from an isothermal vertical square cylinder with exposed upper surface in air mounted on a flat adiabatic base was studied by Oosthuizen [27]. His solution has values of the parameters w/h between 0.1 and 1, Rayleigh number between 1×10^4 and 1×10^7, and Prandtl number of 0.7. His results showed that at the lower values of Rayleigh number and w/hconsidered the mean Nusselt number increases with decreasing w/h, while at larger values of Rayleigh number and w/h the Nusselt number is essentially independent of w/h. Also the mean Nusselt number for the heated top surface was shown to be much lower than that for the vertical side surfaces. He provide a correlation equations for the top and the side surfaces.

Heat transfer from cylinders with a square or rectangular cross-section has mainly been studied from the point of view of their use in pin–fin heat sinks. Typical of these studies are those of Starner and McManus [28], Welling, and Wooldridge [29], Harahap, and McManus [30], Leung, et al. [31], Leung and Probert [32], Yuncu and Anbar [33], Mobedi and Yuncu [34], Harahap, and Rudianto [35], Sahray, et al. [36], Harahap, et al. [37], Yazicioglu, and Yancu [38], and Sahray, et al. [39]. The present study is an extension of that described in Refs. [40–42].

2 EXPERIMENTAL APPARATUS AND PROCEDURES

To study the natural convective heat transfer from a square cylinder which is inclined at an angle to the vertical, a free convection test chamber in which the model could be mounted was designed and constructed. Figure 3 shows a schematic front view of the experimental apparatus. The experimental test chamber had a size of 120 cm height × 25 cm width × 30 cm depth of Plexiglass. It was opened from the top and bottom mounted on a horizontal floor so it surrounds the test section. The volume of the test chamber was large enough to ensure that it did not interfere with the flow over and heat transfer rates from the model. The test chamber was constructed using transparent acrylic plates and constructed in such a way that it can be rotated around a fixed horizontal axis, in this way allowing the inclination angle of the test section to be changed. Furthermore the test chamber is placed in a bigger box 190 cm height × 135 cm width × 125 cm depth which was vented to the room. This arrangement ensures that no external disturbances in the room air or short term temperature changes in the room interfere with the experiments.

The active component of the test apparatus was the solid aluminum square cylinder models used in the experiments, the model sizes are given in Table 1. The square cylinder model has a height of h and cross-sectional width w. The aspect ratio of the models used w/h range from 0.25 to 1. The bottom surface of the model is attached to a large base made of plexiglass 29 cm in length, 23.5 cm in width, and 1.2 cm thick. The end of the model in contact with this base being internally chamfered in order to reduce the contact area between the model and the sheet in order to reduce the conduction heat transfer from the model to the base. A series of small diameter holes were drilled longitudinally to various depths into the models. Seven thermocouples type-T 0.25 mm in diameter were inserted into these holes in various locations were used to measure temperatures. The thermocouples were monitored using a data acquisition system manufactured by TechmaTron Instrument (model USB-TC) which was self calibrating and that in turn was connected to a personal computer. The thermocouples with the data acquisition system were calibrated in a digital temperature controller water bath manufactured by VWR model 1166D, with the calibrated reference thermometer manufactured by Guildline Instruments Company model 9535/01,02,03 between 20 and 100 °C and their output are less than ±0.5 °C of the actual temperature. As noted before the model assemblies could be mounted inside the test chamber as shown in Fig. 3. This allowed the models to be set at any angle to the vertical.

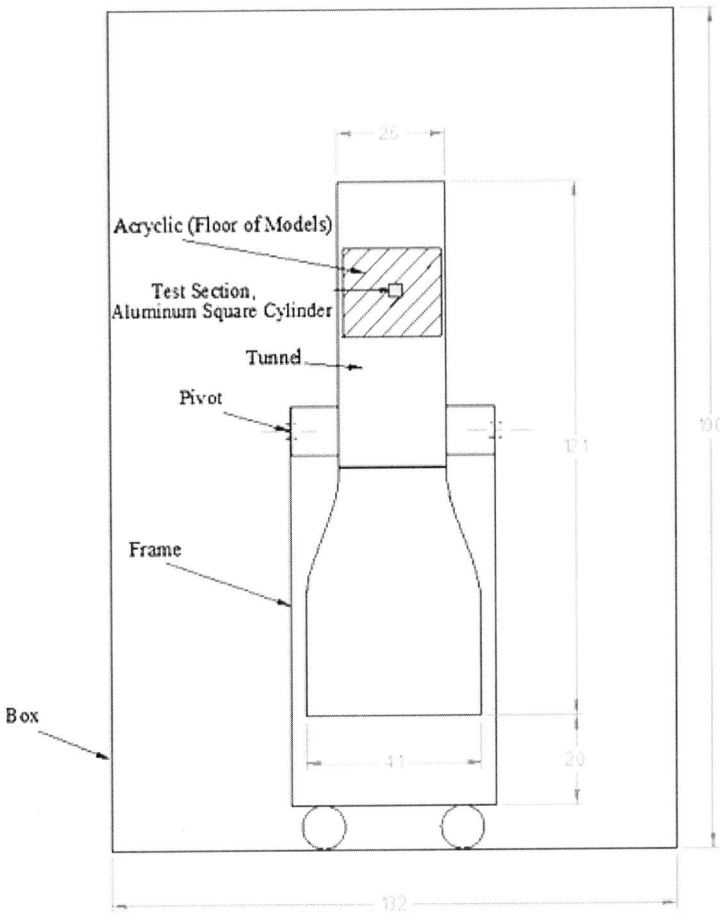

Figure 3. Experimental apparatus, dimensions are in cm

Table 1. Model dimensions

Model no.	Width (w) mm	Height (h) mm	$W = w/h$
1	25.4	25.4	1.0
2	25.4	33.3	0.75
3	25.4	50.2	0.50
4	25.4	100.1	0.25

2.1 Derivation of the experimental results

The heat transfer rates were determined using the transient method, i.e., by heating a model being tested and then measuring its temperature–time variation while it cooled. In an actual test, the model being tested was set at a required angle and then heated in an oven to a temperature of about 140 °C. The model temperature variation with time was then measured while it cooled from 95–40 °C. Tests and approximate calculations indicated that, because the Biot numbers existing during the tests were very small less than 1×10^{-4} which meets the general requirement for applicability of the lumped system method of $Bi < 10^{-1}$. As a result of low Biot number, the temperature of the aluminum models remained effectively uniform at any given instant of time during the cooling process. The overall heat transfer coefficient could then be determined from the measured temperature–time variation using the usual procedure, i.e., the temperature variation of the model was therefore related to the time by:

$$\left(\frac{h_t A}{mC}\right) t = \ln\left(\frac{T_i - T_\infty}{T_e - T_\infty}\right) \tag{1}$$

Hence, h_t could be determined from the measured variation of $ln(T_i - T_\infty)/(T_e - T_\infty)$ with t, and known value of (A/mc).

The value of h_t so determined is, of course, made up of the convective heat transfer to the surrounding air, the radiant heat transfer to the surroundings and the conduction from the model to the base. The radiant heat transfer could be allowed for by calculation using the known emissivity 0.1 of the polished surface of the aluminum models. So the radiant heat transfer coefficient could be calculated as,

$$h_r = \varepsilon\,\sigma\left(T_{W_{avg}} + T_\infty\right)\left(T_{W_{avg}}^2 + T_\infty^2\right) \tag{2}$$

The conduction heat transfer to the base was calculated by fully covering the models with Styrofoam insulation and using the transient method as described before. The convective rate of the heat coefficient by convection is calculated as:

$$h_c = h_t - h_r - h_{cd} \tag{3}$$

The estimated values of convection, conduction and radiation heat transfer coefficients depend on the size of the experimental model, orientation and Rayleigh number, e.g. when the Rayleigh number is equal to 5.89×10^4 and $W = 1$, then $h_c \approx 66.8\ \%$, $h_{cd} \approx 29.9\ \%$ and $h_r < 3.3\ \%$ while when the Rayleigh number is equal to 3.62×10^6 and $W = 0.25$, then $h_c \approx 78$ %, $h_{cd} \approx 17.4$ % and $h_r < 4.6$ % of the total heat transfer.

The dimensional analysis indicates that natural convection heat transfer from short square cylinders inclined at an angle to the vertical depends on Nusselt number, *Nu*, Rayleigh number, *Ra*, the ratio of the width to the height of the square cylinder, *W*,and the inclination angle to the vertical, φ, i.e.;

$$Nu = f(Ra, W, \varphi) \tag{4}$$

where *Ra* is defined as,

$$Ra = \frac{\beta g (T_{W_{avg}} - T_F) h^3}{\nu \alpha} \tag{5}$$

and the average Nusselt number is defined as usual as:

$$Nu = \frac{h_c h}{k} \tag{6}$$

In expressing the experimental results in dimensionless form, all air properties in the Nusselt and Rayleigh numbers have been evaluated at the mean film temperature (Twavg+TF)/2(Twavg+TF)/2 existing during the test.

The uncertainty in the present experimental values of Nusselt number arises due to uncertainties in the temperature measurements, non-uniform heating, temperature differences in the model during cooling, and uncertainties in the corrections applied for conduction heat transfer through the base. Therefore, an uncertainty analysis was performed by applying the estimation method proposed by Moffat [43, 44]. The uncertainty in the average Nusselt number was found to be a function of temperature, the angle of inclination and the size of the experimental model used. Overall, the uncertainty in the average Nusselt number was estimated to be less than ±12 % for selected time interval, and the scatter in the experimental data is between ±3 % of the mean values.

Tests were performed with models mounted at various angles of inclination between vertically upwards and vertically downwards as shown Fig. 2. This work will focus on the effect of inclination angle and the aspect ratio of the square cylinder with exposed top surface on the mean heat transfer rate.

3 NUMERICAL SOLUTION PROCEDURE

The solution domain used in obtaining the numerical solution is shown in Fig. 4. The flow has been assumed to be symmetric about the vertical center plane DEFOTSIND shown in Fig. 4 and to be steady and laminar. It has also been assumed that the fluid properties are constant except for the density change with temperature which gives rise to the buoyancy forces, this being treated here by using the Boussinesq approach. The governing equations have been written in dimensionless form using the height, *h*, of the cylinder as the length scale and $T_W - T_F$ as the temperature scale, T_W being the temperature of surface of cylinder and T_F being the fluid temperature far from the cylinder. The following reference velocity has been introduced:

$$u_r = \frac{\alpha}{h}\sqrt{Ra\,Pr} \tag{7}$$

where *Pr* is the Prandtl number and *Ra* is the Rayleigh number based on *h*, i.e.:

$$Ra = \frac{\beta g (T_W - T_F) h^3}{\nu \alpha} \tag{8}$$

The following dimensionless variables have then been defined:

$$X = \frac{x}{h}, Y = \frac{y}{h}, Z = \frac{z}{h}, U_X = \frac{u_x}{u_r},$$
$$U_Y = \frac{u_y}{u_r}, U_Z = \frac{u_z}{u_r}, P = \frac{(p - p_F)h}{\mu u_r}, \tag{9}$$
$$\theta = \frac{T - T_F}{T_W - T_F}$$

where *T* is the temperature, T_W is the temperature of the hot cylinder walls and T_F is the fluid temperature far from the cylinder. As shown in Fig. 4, the *z*-coordinate is measured along the axis of the cylinder and the *x*- and *y*-coordinate are measured in the plane of the flat adiabatic base plate. In terms of these dimensionless variables and if φ is the angle from the

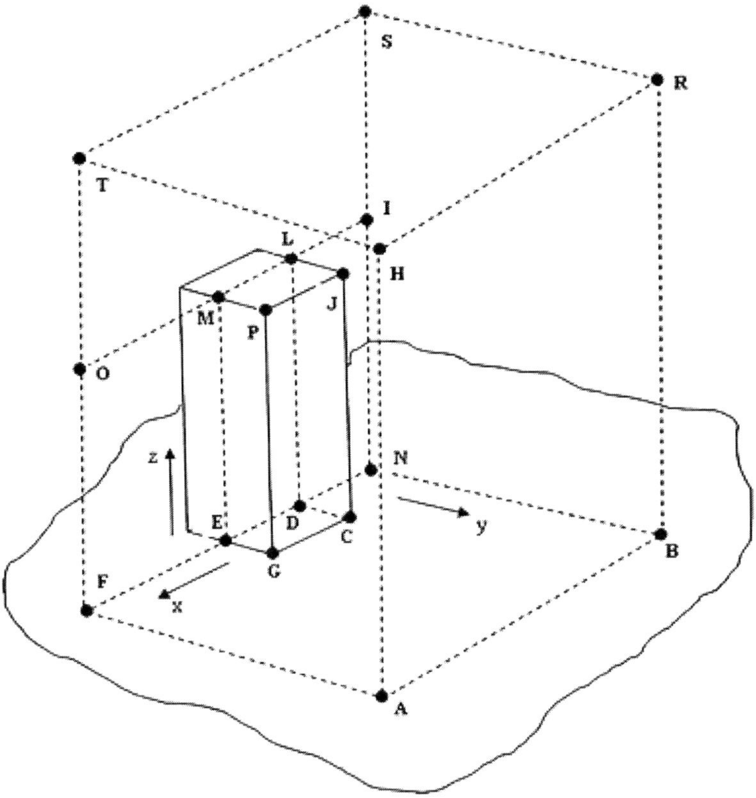

Figure 4. Solution domain

vertical at which the plate is inclined the governing equations are the following:

$$\frac{\partial U_X}{\partial X} + \frac{\partial U_Y}{\partial Y} + \frac{\partial U_Z}{\partial Z} = 0 \tag{10}$$

$$U_X \frac{\partial U_X}{\partial X} + U_Y \frac{\partial U_X}{\partial Y} + U_Z \frac{\partial U_X}{\partial Z} = \sqrt{\frac{Pr}{Ra}} \left(-\frac{\partial P}{\partial X} + \frac{\partial^2 U_X}{\partial X^2} + \frac{\partial^2 U_X}{\partial Y^2} + \frac{\partial^2 U_X}{\partial Z^2} \right) + \theta \sin \varphi \tag{11}$$

$$U_X \frac{\partial U_Y}{\partial X} + U_Y \frac{\partial U_Y}{\partial X} + U_Z \frac{\partial U_Y}{\partial X} = \sqrt{\frac{Pr}{Ra}} \left(-\frac{\partial P}{\partial Y} + \frac{\partial^2 U_Y}{\partial X^2} + \frac{\partial^2 U_Y}{\partial Y^2} + \frac{\partial^2 U_Y}{\partial Z^2} \right) \quad (12)$$

$$U_X \frac{\partial U_Z}{\partial X} + U_Y \frac{\partial U_Z}{\partial Y} + U_Z \frac{\partial U_Z}{\partial Z} = \sqrt{\frac{Pr}{Ra}} \left(-\frac{\partial P}{\partial Z} + \frac{\partial^2 U_Z}{\partial X^2} + \frac{\partial^2 U_Z}{\partial Y^2} + \frac{\partial^2 U_Z}{\partial Z^2} \right) + \theta \cos \varphi \quad (13)$$

$$U_X \frac{\partial \theta}{\partial X} + U_Y \frac{\partial \theta}{\partial Y} + U_Z \frac{\partial \theta}{\partial Z} = \frac{1}{\sqrt{Ra Pr}} \left(\frac{\partial^2 \theta}{\partial X^2} + \frac{\partial^2 \theta}{\partial Y^2} + \frac{\partial^2 \theta}{\partial Z^2} \right) \quad (14)$$

Because the flow has been assumed to be symmetric about the vertical center-plane of the cylinder, the domain used in obtaining the solution is as shown in Fig. 4. The assumed boundary conditions on the solution in terms of the dimensionless variables are that for all the cylinder surfaces the dimensionless velocity components are all equal to zero and the dimensionless temperature is equal 1. On the adiabatic base the assumed boundary conditions are that all dimensionless velocities are equal to zero and that the dimensionless temperature gradient normal to all surfaces is equal to 0. On the plane of symmetry the dimensionless velocity component normal to the plane is zero and the gradients normal to this plane of the remaining velocity components and temperature are zero. On the outer planes of the solution domain the dimensionless temperature is equal to zero and the velocity components in the plane parallel to the surface are zero. On the upper surface HTSR the boundary conditions are the dimensionless temperature is equal to zero and the pressure is equal to zero. The outer boundary conditions of the solution domain were changed as the domain rotated from 0° to 180°.

The above dimensionless governing equations subject to the boundary conditions discussed above have been numerically solved using the commercial finite-volume solver, FLUENT. In obtaining the numerical results the pressure-based solver was used with the second-order upwind scheme for convective terms in the mass, momentum and energy equations. For pressure discretization, the Presto scheme has been employed while the SIMPLE-algorithm has been used for pressure–velocity coupling discretization. Extensive grid and convergence criterion independence testing was undertaken for all cases studied in the present work. A non-uniform grid distribution is used in the plane perpendicular and parallel to the main flow direction. This indicated that the heat transfer results presented here are to less than 1 % independent of the number of grid points and of the convergence-criterion used. Some typical results obtained here in the grid-independence testing are shown in Tables 2 and 3. These results are for the case of natural convective heat transfer from vertical and inclined square cylinders when dimensionless width,W, is equal to 1. In obtaining the results the number of nodes used was 1,191,641. Close to the square cylinder, the number of grid points or control volumes is increased to enhance the resolution and accuracy. The results in Tables 2 and 3show that the mean Nusselt number for the surfaces that make up the cylinder for the selected number of nodes are to less than 1 % independent of the number of grid points. The effect of the distances of the outer surfaces of the solution domain (i.e., surfaces BRSN, AHTOF, ABRH, and HRST in Fig. 4) from the heated surfaces was also examined and the positions used in obtaining the results discussed here were chosen to ensure that the heat transfer results were independent of this positioning to within 1 %.

Table 2. Results for the mesh independence study for $W = 1$, $Ra = 1E6$ and $\varphi = 0°$, 90° and 180°

Rayleigh no.	Tilt angle	No. of nodes	Nu_{left}	Nu_{right}	Nu_{Front}	Nu_{Top}	Nu_{Total}
1×10^6	0°	532,408	17.485	17.57	17.633	6.11	15.298
		628,730	17.284	17.344	17.536	6.036	15.147
		764,791	17.117	17.159	17.575	6	15.085
		837,385	17.026	17.072	17.223	5.932	14.895
		930,109	16.933	16.99	17.116	5.955	14.822
		1,040,126	16.648	16.648	16.578	5.85	14.46
		1,191,641	16.607	16.607	16.545	5.871	14.435
		1,223,280	16.595	16.595	16.54	5.88	14.43
1×10^6	90°	930,109	6.288	12.446	15.327	15.955	13.068
		1,040,126	6.826	12.231	14.913	16.389	13.054
		1,191,641	6.877	12.184	14.891	16.397	13.048
		1,223,280	6.895	12.178	14.892	16.398	13.051
1×10^6	180°	930,109	14.195	14.254	14.246	14.102	14.21
		1,040,126	14.076	14.076	14.033	14.228	14.09
		1,191,641	14.054	14.054	14.004	14.127	14.05
		1,223,280	14.063	14.063	14.028	14.183	14.07

Table 3. Results for the mesh independence study for $W = 1$, $Ra = 1E4$ and $\varphi = 0°, 90°, 180°$

Rayleigh no.	Tilt angle	No. of nodes	Nu_{left}	Nu_{right}	Nu_{Front}	Nu_{Top}	Nu_{Total}
1×10^4	0°	930,109	5.365	5.363	5.296	2.513	4.766
		1,040,126	5.339	5.339	5.353	2.486	4.774
		1,191,641	5.342	5.342	5.355	2.485	4.776
		1,223,280	5.346	5.346	5.352	2.485	4.777
1×10^4	90°	930,109	2.01	5.122	5.036	5.567	4.554
		1,040,126	2.027	5.101	5.023	5.5106	4.537
		1,191,641	2.036	5.1103	5.046	5.484	4.544
		1,223,280	2.032	5.102	5.023	5.5105	4.538
1×10^4	180°	930,109	4.362	4.357	4.326	5.934	4.661
		1,040,126	4.382	4.382	4.382	5.871	4.68
		1,191,641	4.376	4.376	4.375	5.825	4.666
		1,223,280	4.389	4.389	4.383	5.863	4.682

The mean Nusselt number for the heated cylinder is defined by:

$$Nu = \frac{\bar{q}' h}{k(T_w - T_F)} \tag{15}$$

where q¯′q¯′ is the mean heat transfer rate per unit area from the entire surface of the cylinder. Similarly the mean Nusselt numbers for the left, right, front and top side surfaces of the cylinder are defined as follows:

$$Nu_{left} = \frac{q'_{left} h}{k(T_w - T_F)} \tag{16}$$

$$Nu_{right} = \frac{q'_{right} h}{k(T_w - T_F)} \tag{17}$$

$$Nu_{front} = \frac{q'_{front} h}{k(T_w - T_F)} \tag{18}$$

$$Nu_{Top} = \frac{q'_{top} h}{k(T_W - T_F)} \quad (19)$$

where q'leftqleft', q'rightqright', q'frontqfront' and q'topqtop' are the mean heat transfer rates per unit area from the left, right, front and top side surfaces of the cylinder respectively.

4 RESULTS

The solution has the following parameters:

1. The Rayleigh number, *Ra*, based on the 'height' of the heated cylinder, *h*, and the overall temperature difference $T_W - T_F$
2. The dimensionless width of the cylinder, $W = w/h$
3. The Prandtl number, *Pr*
4. The inclination angle φ of the cylinder

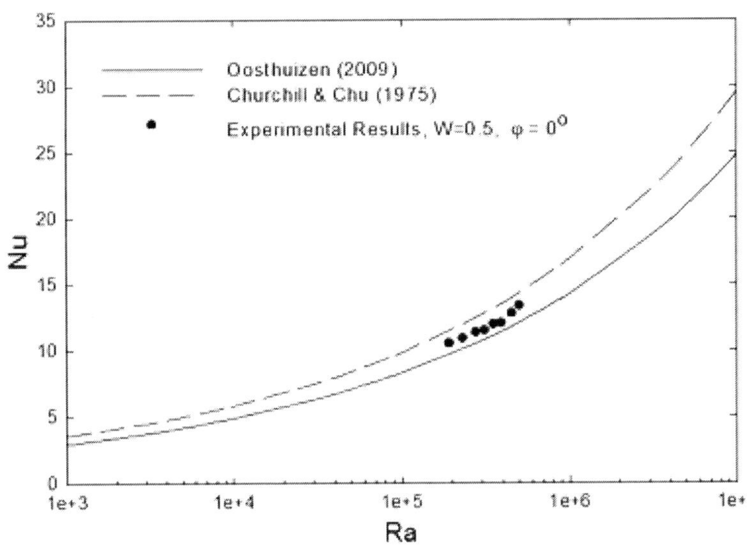

Figure 5. Comparison of mean Nusselt number values given by correlation equations and experimental results for $W = 0.5$ and $\varphi = 0°$

Because of the applications that motivated this study, results have only been obtained for $Pr = 0.7$. A wide range of the other governing parameters have been considered.

In order to validate the results a comparison between the present experimental results and the correlation equations by Churchill and Chu [45] and Oosthuizen [27] are shown in Fig. 5. Oosthuizen [27] correlation equation is valid for a vertical isothermal square cylinder with an exposed top surface where the dimensionless width of the cylinder is less than 0.6. Churchill and Chu [45] correlation equation are valid for a wide plate. Good agreement between the correlation equations and the experimental results for isothermal vertical square cylinder for $W = 0.5$ where the heat transfer from the top surface is small compared to the heat transfer from the side surfaces.

Typical variations of numerical and experimental values results of the mean Nusselt number for the entire cylinder, Nu, with Rayleigh number, Ra, for various values of the dimensionless cylinder width, W, at different angles of inclination, φ, between vertically upward and vertically downward are shown in Figs. 6, 7, 8, 9 and 10. These figure s show that the mean Nusselt number increases as the Rayleigh number increases and the dimensionless cylinder width decreases for all values of inclination angles considered in this study. The increase in the Nusselt number with decreasing dimensionless cylinder width, W, arises from the fact that there is an induced inflow from the edges towards the center of the faces of the square cylinder and this causes the heat transfer rate to be higher near the vertical edges of the square cylinder than it is in the center region of the cylinder. The experimental results given in these Figure s seen to be in good agreement with the numerical results and shows the same trends as the numerical results. However it will be seen from Figs. 9 and 10 that for angle of inclination of 135° and 180° the experimental results have lower values than the numerical results. This is because as the angle of inclination increases above 90° the base of the square cylinder which is made of plexiglass as mentioned before gets heated by the upward buoyant flow over the square cylinder and can no longer be treated as adiabatic as assumed in the numerical work. Experimental results in the dimensionless form for different dimensionless width, W, and different angles of inclination are given in Tables 4, 5, 6and 7.

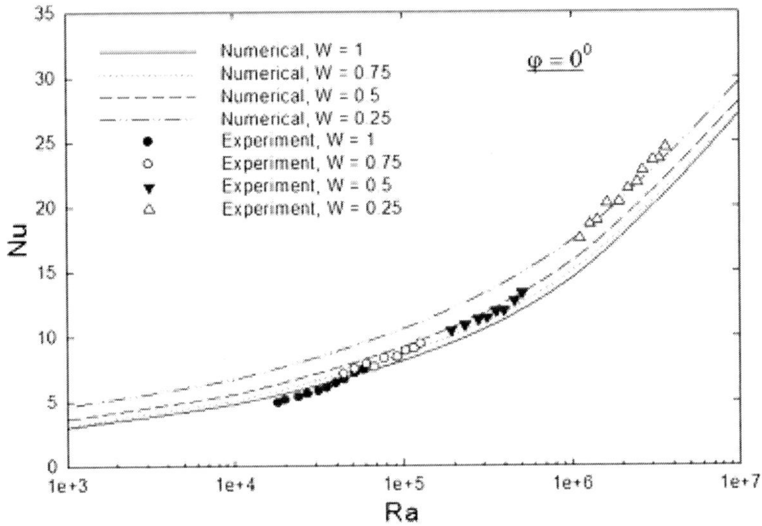

Figure 6. Variation of mean Nusselt number for the square cylinder with Rayleigh number for various values of dimensionless cylinder width, *W*, when $\varphi = 0°$

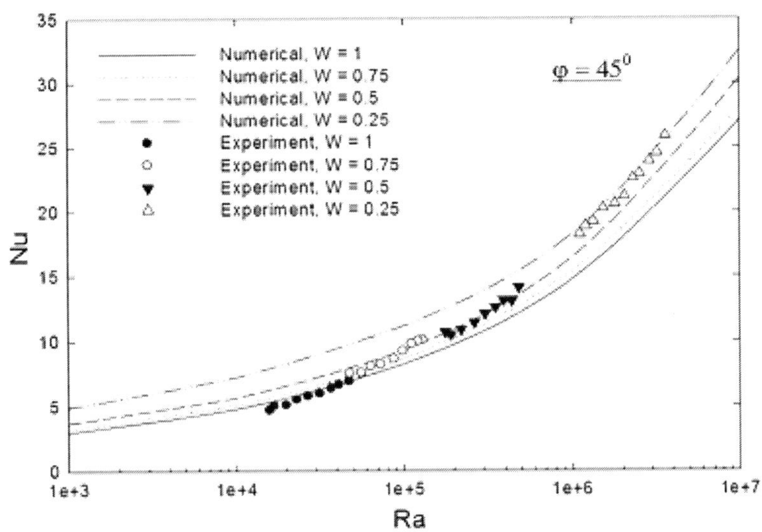

Figure 7. Variation of mean Nusselt number for the square cylinder with Rayleigh number for various values of dimensionless cylinder width, *W*, when $\varphi = 45°$

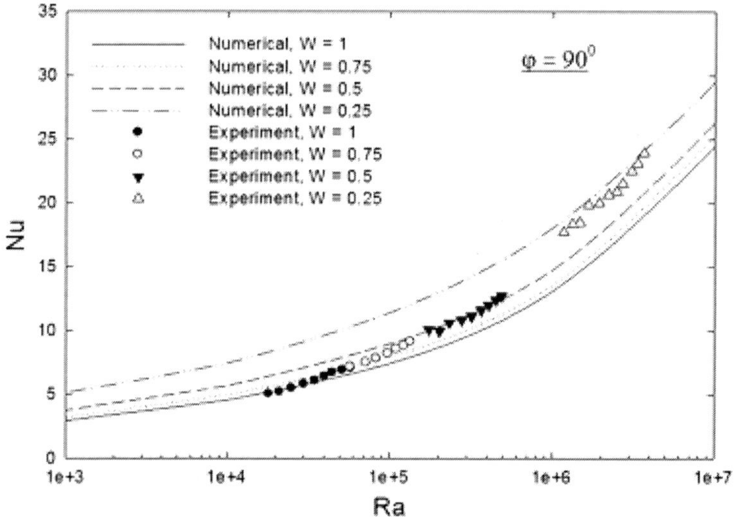

Figure 8. Variation of mean Nusselt number for the square cylinder with Rayleigh number for various values of dimensionless cylinder width, *W*, when $\varphi = 90°$

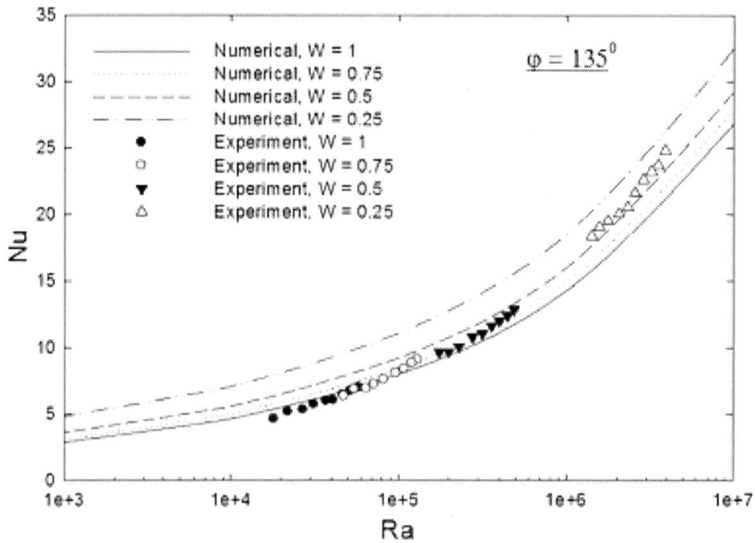

Figuer 9. Variation of mean Nusselt number for the square cylinder with Rayleigh number for various values of dimensionless cylinder width, *W*, when $\varphi = 135°$

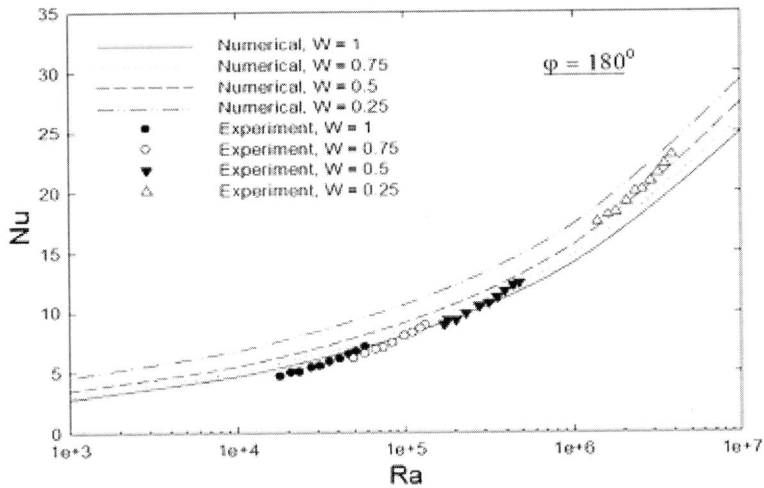

Figure 10. Variation of mean Nusselt number for the square cylinder with Rayleigh number for various values of dimensionless cylinder width, *W*, when $\varphi = 180°$

Table 4. Experimental results for $W = 1$

$\varphi = 0°$		$\varphi = 45°$		$\varphi = 90°$		$\varphi = 135°$		$\varphi = 180°$	
Ra	Nu	Ra	Nu	Ra	Nu	Ra	Nu	Ra	Nu
5.887E+04	7.37	5.667E+04	7.456	5.826E+04	7.062	5.863E+04	7.003	5.840E+04	7.061
5.224E+04	7.155	4.747E+04	6.9	5.163E+04	6.895	5.192E+04	6.722	5.150E+04	6.693
4.508E+04	6.67	4.124E+04	6.63	4.504E+04	6.72	4.680E+04	6.448	4.670E+04	6.464
4.022E+04	6.39	3.708E+04	6.36	4.020E+04	6.39	4.115E+04	6.04	4.120E+04	6.08
3.569E+04	6.034	3.169E+04	5.98	3.510E+04	6.056	3.715E+04	5.965	3.580E+04	5.825
3.184E+04	5.79	2.710E+04	5.77	3.004E+04	5.81	3.151E+04	5.693	3.130E+04	5.5
2.730E+04	5.544	2.323E+04	5.482	2.530E+04	5.48	2.714E+04	5.31	2.780E+04	5.353
2.420E+04	5.317	2.009E+04	5.07	2.137E+04	5.18	2.211E+04	5.153	2.350E+04	5
2.015E+04	5.09	1.703E+04	5	1.830E+04	5.03	1.828E+04	4.61	2.100E+04	4.951
1.830E+04	4.88	1.594E+04	4.66	–	–	–	–	1.810E+04	4.641

Table 5. Experimental results for $W = 0.75$

$\varphi = 0°$		$\varphi = 45°$		$\varphi = 90°$		$\varphi = 135°$		$\varphi = 180°$	
Ra	Nu	Ra	Nu	Ra	Nu	Ra	Nu	Ra	Nu
1.281E+05	9.39	1.303E+05	10.034	1.350E+05	9.121	1.304E+05	9.071	1.340E+05	8.844
1.176E+05	8.994	1.227E+05	9.951	1.231E+05	8.787	1.211E+05	8.837	1.247E+05	8.565
1.050E+05	8.834	1.121E+05	9.745	1.107E+05	8.55	1.086E+05	8.374	1.120E+05	8.217
9.330E+04	8.353	9.955E+04	9.188	9.852E+04	8.186	9.701E+04	8.072	1.003E+05	7.933
7.834E+04	8.284	8.801E+04	8.585	8.340E+04	7.83	8.211E+04	7.59	8.526E+04	7.37
6.834E+04	7.575	7.350E+04	8.163	7.205E+04	7.51	7.210E+04	7.23	7.500E+04	6.973
6.109E+04	7.832	6.387E+04	8.02	5.780E+04	7.155	6.480E+04	6.922	6.766E+04	6.833
5.211E+04	7.455	5.618E+04	7.54	–	–	5.500E+04	6.851	5.820E+04	6.5
4.471E+04	7.065	4.783E+04	7.51	–	–	4.756E+04	6.344	4.970E+04	6.09

Table 6. Experimental results for $W = 0.5$

φ = 0°		φ = 45°		φ = 90°		φ = 135°		φ = 180°	
Ra	Nu	Ra	Nu	Ra	Nu	Ra	Nu	Ra	Nu
5.067E+05	13.292	4.830E+05	14.091	4.877E+05	12.684	4.866E+05	12.843	4.798E+05	12.395
4.570E+05	12.74	4.380E+05	13.086	4.491E+05	12.39	4.453E+05	12.4114	4.418E+05	12.194
3.950E+05	11.983	3.917E+05	13.107	4.056E+05	12	3.975E+05	12.024	3.933E+05	11.721
3.547E+05	11.905	3.508E+05	12.59	3.646E+05	11.582	3.568E+05	11.647	3.543E+05	11.205
3.142E+05	11.435	3.045E+05	12.073	3.184E+05	11.12	3.143E+05	11.065	3.152E+05	10.721
2.790E+05	11.286	2.633E+05	11.404	2.771E+05	10.821	2.751E+05	10.815	2.766E+05	10.44
2.332E+05	10.842	2.198E+05	10.864	2.322E+05	10.611	2.285E+05	10.1	2.310E+05	9.87
1.936E+05	10.47	1.906E+05	10.467	2.024E+05	9.944	1.983E+05	9.71	2.014E+05	9.272
–	–	1.772E+05	10.69	1.745E+05	10.064	1.747E+05	9.672	1.808E+05	9.312
–	–	–	–	–	–	–	–	1.715E+05	8.931

Table 7. Experimental results for $W = 0.25$

φ = 0°		φ = 45°		φ = 90°		φ = 135°		φ = 180°	
Ra	Nu	Ra	Nu	Ra	Nu	Ra	Nu	Ra	Nu
3.617E+06	24.512	3.611E+06	25.843	3.663E+06	23.89	3.873E+06	24.8	3.856E+06	23.04
3.331E+06	23.611	3.218E+06	24.563	3.333E+06	23.082	3.524E+06	23.7	3.497E+06	22.27
3.050E+06	23.531	2.904E+06	23.894	3.064E+06	22.5	3.212E+06	23.244	3.201E+06	21.55
2.649E+06	22.744	2.533E+06	22.923	2.693E+06	21.524	2.880E+06	22.61	2.890E+06	20.85
2.460E+06	21.835	2.306E+06	22.544	2.478E+06	20.88	2.560E+06	21.612	2.580E+06	20.18
2.165E+06	21.315	2.052E+06	21.16	2.201E+06	20.631	2.324E+06	20.541	2.350E+06	20
1.914E+06	20.334	1.796E+06	20.58	1.936E+06	19.97	2.050E+06	20.06	2.079E+06	19.19
1.617E+06	20.226	1.527E+06	20.21	1.659E+06	19.825	1.760E+06	19.516	1.796E+06	18.29
1.412E+06	18.93	1.335E+06	19.2	1.466E+06	18.462	1.564E+06	19.06	1.612E+06	18.05
1.276E+06	18.6	1.202E+06	18.86	1.326E+06	18.326	1.408E+06	18.39	1.391E+06	17.475

Typical variations of the mean Nusselt number for the entire cylinder, *Nu*, with dimensionless cylinder width, *W*, for various values of *Ra* and φ are shown in Fig. 11. It will be seen from this Figure that, the mean Nusselt number increases with decreasing *W* essentially at the lower values of *W* considered. At the lower values of *Ra* considered and when the dimensionless cylinder width, *W*, approximately greater than 0.6 the mean Nusselt number is essentially independent of the angle of inclination, φ. However at larger values of *Ra* the mean Nusselt number depends on the angle of inclination φ.

Figure s 12 and 13 shows the variation of the mean Nusselt number with angle of inclination for different values of *W*. Figure 12 shows a good agreement between the numerical and the experimental results for inclination angles below 90° and that the experimental results have lower values than the numerical results as the angle of inclination gets larger than 90° as discussed before. Considering the numerical results given in Fig. 13 which shows the variation of the mean Nusselt number for the

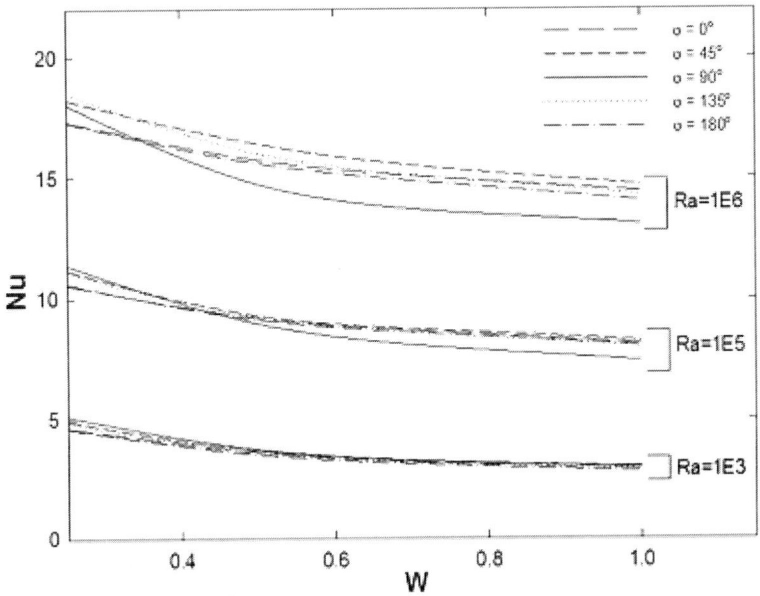

Figure 11. Variation of mean Nusselt number for the cylinder with dimensionless cylinder width for various values of Rayleigh number and φ

square cylinder with φ for various values of Rayleigh number and of the dimensionless cylinder width, W, it will clearly be seen that the mean Nusselt number at the lower values of Rayleigh numbers varies differently than at the higher values of the Rayleigh numbers. As already mentioned at the lower values of the Rayleigh number the mean Nusselt number is almost independent of the angle of inclination. However at higher values of Rayleigh number this is not the case. At the higher values of the Rayleigh number when dimensionless square cylinder width, W, is bigger than 0.25 the lowest mean Nusselt number occurs when the cylinder is in a horizontal position, i.e., when φ is equal 90° while the highest mean Nusselt number occurs when the cylinder is at inclination angles, φ, of 45° and 135°. When the dimensionless square cylinder width, W, is equal to 0.25 the lowest mean Nusselt number occurs when the cylinder is in a vertical position, i.e., when φ is equal 0° and 180° while the highest mean Nusselt number occurs when the cylinder is at inclination angles, φ, of between 45° and 135°.

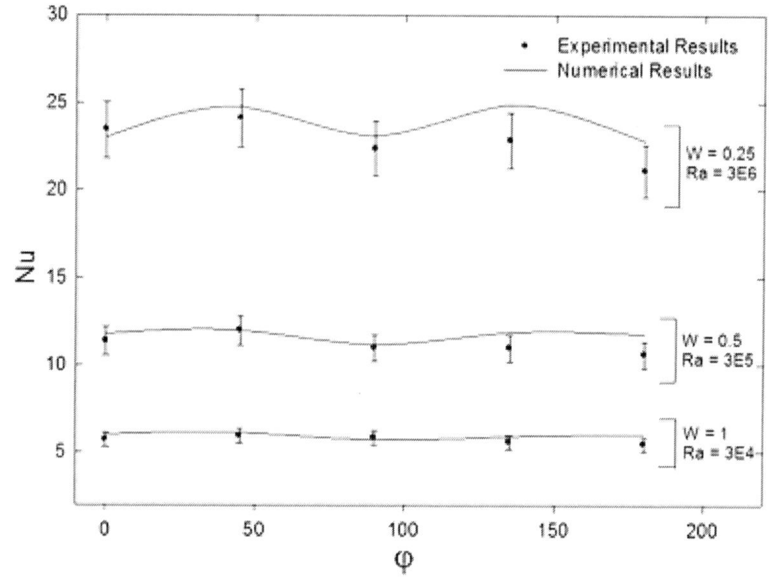

Figure 12. Variation of mean Nusselt number for the cylinder with φ for various values of Rayleigh number and dimensionless cylinder width, W

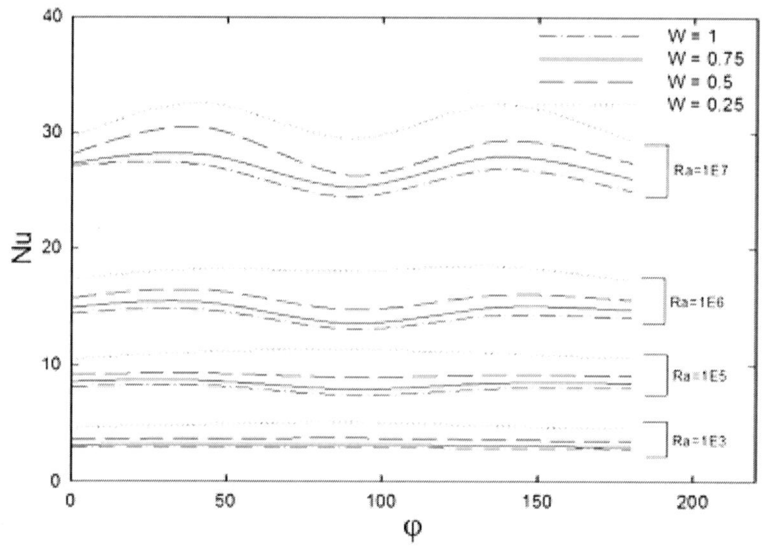

Figure 13. Variation of mean Nusselt number for the cylinder with φ for various values of Rayleigh number and dimensionless cylinder width W

In some situations the mean heat transfer rates from the individual surfaces, i.e., the left side, the right side, the front, and the top surfaces of the cylinder are of importance and these will be considered next. Typical variations of the mean Nusselt numbers for the front, right side, left side, and top surfaces with angles of inclination of various values of W and Ra are shown in Figs. 14, 15, 16 and 17, respectively.

Figure 14 shows that Nu_{front} increases as the dimensionless cylinder width Wdecreases and that the variation of Nu_{front} with φ at lower values of Rayleigh number differs from the variation at higher values of Rayleigh number. At lower values of Rayleigh number Nu_{front} has a maximum value when the inclination angle φ is equal to 90° and a minimum value when φ is equal to 180° whereas at higher values of Rayleigh number the maximum value of Nu_{front} occurs when the angle of inclinationφ is equal to 45° while the lowest value again occurs when the angle of inclination $\varphi$is equal to 180°. It will be noted that at the highest value of Ra considered in Fig. 14,Nu_{front} increases as φ increases from 0° to 45°. Nu_{front} then decreases, as φincreases from 45° to 90°, then increases as φ increases from 90° to 135°, and then again decreases as φ increases from 135° to 180°. When dimensionless square cylinder width, W, is equal to 0.25 the maximum mean Nusselt number occurs when the cylinder is in a horizontal position, i.e., when φ is equal 90°. The results given in Fig. 15 shows that the variation of Nu_{right} with φ exhibits the same basic form of behavior as Nu_{front} (Fig. 14), except that at the highest Rayleigh number considered when dimensionless square cylinder width,W, is equal to 0.25 the maximum mean Nusselt number occurs when φ is equal 45° and 135°.

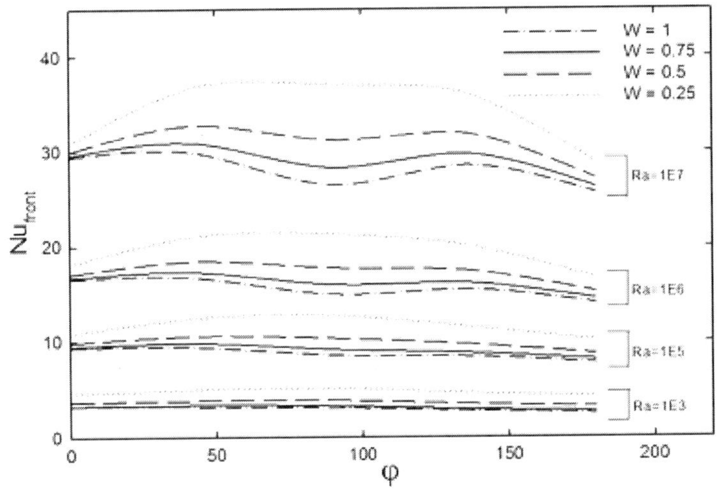

Figure 14. Variation of mean Nusselt number for *front* side heated surface of the cylinder with φ for various values of Rayleigh number and dimensionless cylinder width W

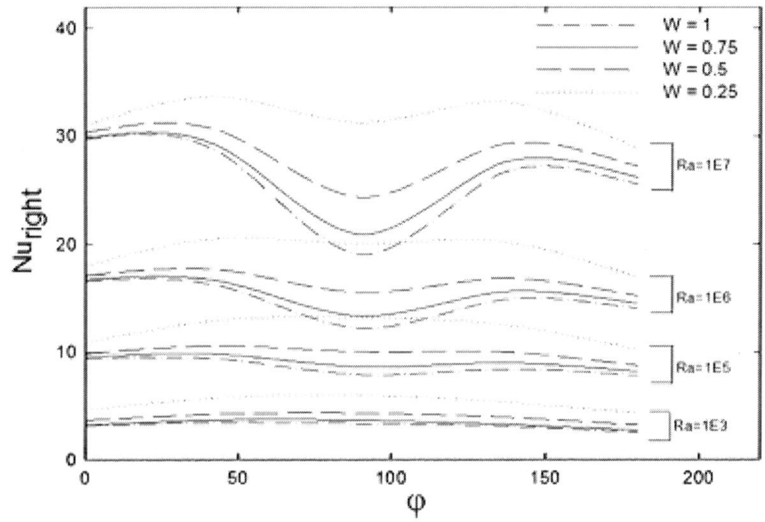

Figure 15. Variation of mean Nusselt number for *left* side heated surface of the cylinder with φ for various values of Rayleigh number and dimensionless cylinder width W

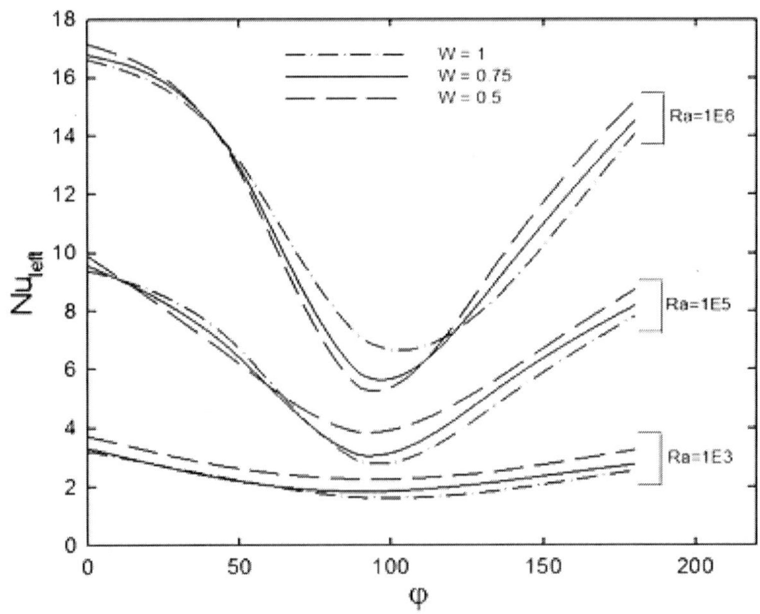

Figure 16. Variation of mean Nusselt number for the *left* heated surface of the cylinder with φ for various values of the Rayleigh number and of the dimensionless cylinder width W

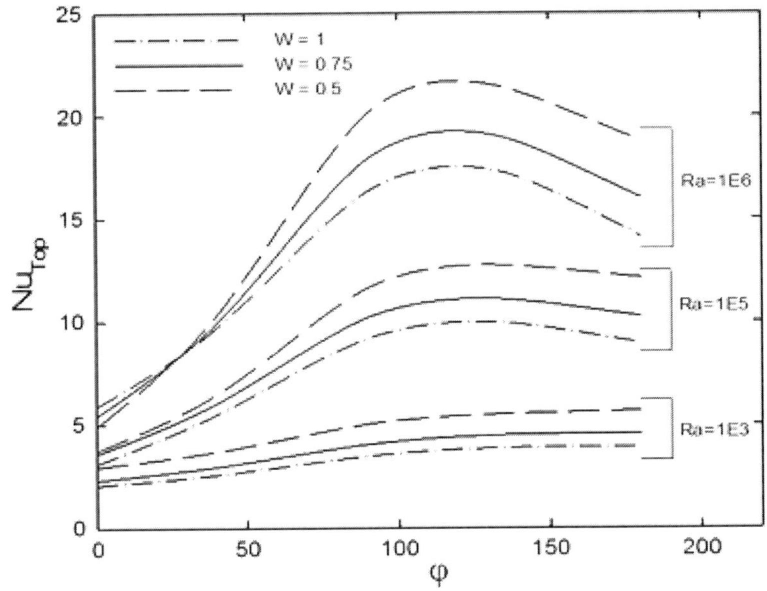

Figure 17. Variation of mean Nusselt number for *top* side heated surface of the cylinder with φ for various values of Rayleigh number and dimensionless cylinder width W

Figure 16 shows the variation of Nu_{left} with φ for various values of Ra and W. It will be seen that at lower Rayleigh numbers Nu_{left} increases as the dimensionless plate width W decreases at all angles of inclination. However at higher Rayleigh numbers a different form of variation of Nu_{left} with W exists. At high Rayleigh numbers Nu_{left} increases as the dimensionless plate width W decreases for values of φ near 0° and 180° but decreases as the dimensionless plate width W decreases for values of φ near 90° This is because as the angle of inclination of the cylinder increases 0° the left side of the cylinder changes from being an effective vertical flat plate to being horizontal plate for φ near 90° and then again become an effective vertical flat plate for φ near 180°.

The changes in the Nusselt number of the individual surfaces is further illustrated by the results given in Fig. 17 which shows that the Nu_{Top} increases as the dimensionless cylinder width W decreases at all angles of inclination except at the highest Rayleigh number considered when at angles of inclination φ near to 0° where Nu_{Top} increases as the dimensionless plate width W increases. It should also be noted that at the lower values of Rayleigh number considered Nu_{Top} increases continuously as the angle of inclination increases. However at higher values of Rayleigh number Nu_{Top} increases as the angle of inclination increases but then passes through a maximum at an angle of inclination φ of about 120° and then decreases with a further increase in φ.

Typical variations of Nu_{Top}, Nu_{front}, Nu_{right}, and Nu_{left}, with an angle of inclination $\varphi$and for two values of Ra and W equal 0.5 are shown in Figs. 18 and 19. Nu_{front}, Nu_{right}, and Nu_{left}, have the same values when the angle of inclination is equal to 0° and to 180°. As expected, it will be seen that Nu_{Top} is much less than Nu_{front}, Nu_{right}, and Nu_{left} when the angle of inclination is equal to 0°. However as the angle of inclination increases this difference decreases and for angles greater than about 75° the Nu_{Top} is greater than Nu_{front}, Nu_{right}, and Nu_{left}. These changes arise from the fact that as the angle of inclination of each of the surfaces relative to the vertical changes, e.g., when the angle of inclination of the cylinder is 0° the 'top' surface faces upwards whereas when the angle of inclination of the cylinder is 180° the 'top' surface faces downwards.

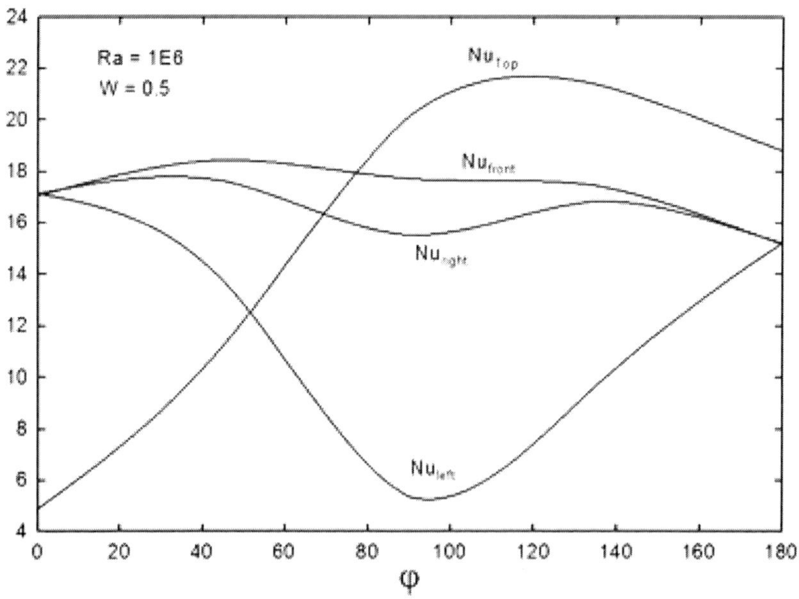

Figure 18. Variation of mean Nusselt number for surfaces that make up the cylinder, *top, front, right* and *left* with φ forRa = 10^6 and for W = 0.5

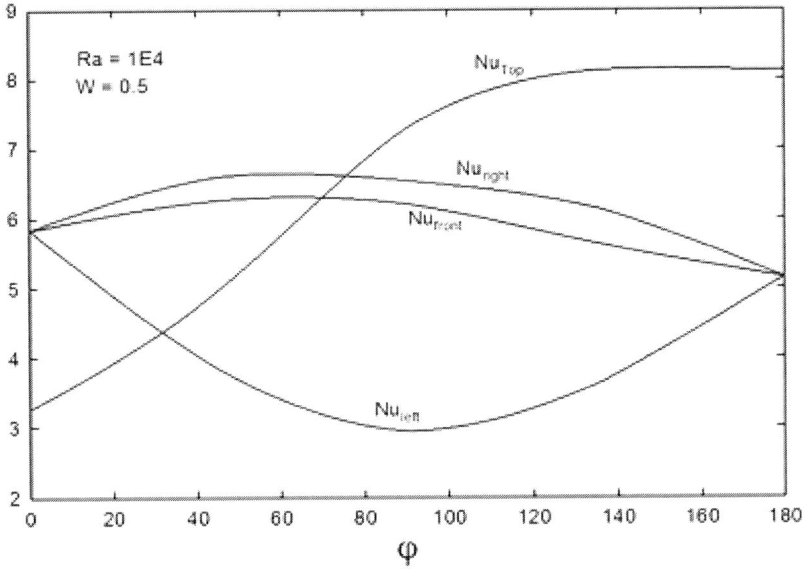

Figure 19. Variation of mean Nusselt number for surfaces that make up the cylinder, *top, front, right* and *left* with φ forRa = 10^4 and for W = 0.5

The changes in the three-dimensional flow patterns and the flow interaction between the surfaces that make up the square cylinder with angle of inclination on the top surface of the square cylinder are illustrated by the local Nusselt number distributions over the heated plate as shown in Fig. 20, where the dimensionless local Nusselt number are defined as:

$$Nu_{L(Top)} = \frac{q'_{Top}\, h}{k(T_w - T_F)} \tag{20}$$

This Figure shows the local Nusselt number distributions over the top surface of the square cylinder for W equal to 0.5 and two values of Rayleigh numbers for different angles of inclination. The results given in this Figure illustrate how the angle of inclination and the interaction of flow at different Rayleigh numbers have an effect on the top surface as the result of interaction of flow over the surfaces that make up the square cylinder and how they significantly have different behavior on the top surface as the Rayleigh numbers and angle of inclination changes which gives rise to a complex flow over the top surface. This Figure shows regions of high and low local Nusselt number for different inclination angle and different Rayleigh numbers.

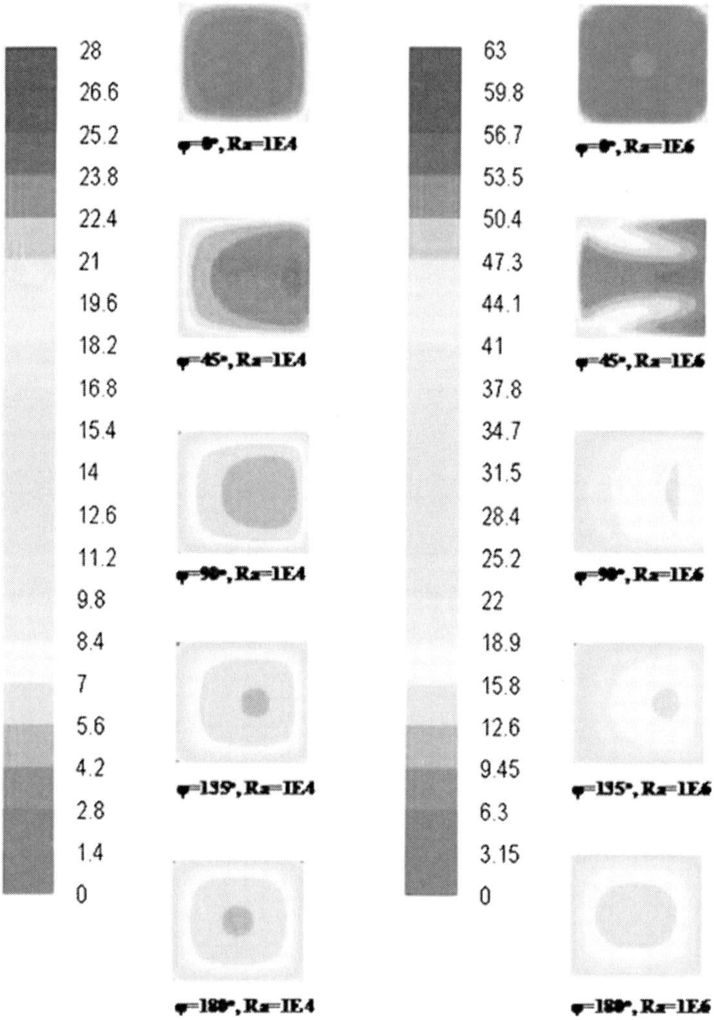

Figure 20. Variation of local Nusselt number based on *h* over *top* side of the square cylinder with different angle of inclinations for *Ra* = 1E4, *Ra* = 1E6 and *W* = 0.5

The variation of the mean Nusselt number for the top surface relative to the mean Nusselt numbers for the other surfaces has already been discussed. The relative importance of the heat transfer rate from the top surface compared to that from the vertical side surfaces of the cylinder will however depend both on the mean Nusselt numbers for the surfaces and the relative surface areas. Now since:

$$A_{side} = 4W, \quad A_{Top} = W^2, \quad \text{and } A_{total} = 4W + W^2 \tag{21}$$

where A_{total}, A_{side}, and A_{Top} are the dimensionless surface areas of the entire cylinder, of all of the 'sides' of the cylinder, and of the 'top' surface of the cylinder respectively, it follows that:

$$\frac{A_{side}}{A_{total}} = \frac{4}{4+W}, \quad \frac{A_{Top}}{A_{total}} = \frac{W}{4+W}, \quad \frac{A_{Top}}{A_{side}} = \frac{W}{4} \tag{22}$$

This equation indicates that over the range of values of W considered here, i.e., 0.25 to 1, A_{Top}/A_{side} is relatively small having a maximum value of 0.25 when $W = 1$, i.e., the area of the top surface relative to the area of the vertical side surfaces remains comparatively small.

Now:

$$\frac{Q_{total}}{Q_{Top}} = \frac{q' A_{total}}{q'_{Top} A_{Top}} = \frac{Nu A_{total}}{Nu_{Top} A_{Top}} = \frac{Nu}{Nu_{Top}} \frac{4+W}{W} \tag{23}$$

This equation allows Q_{total}/Q_{Top} to be found using the calculated Nusselt number values. Some typical results obtained are shown in Fig. 21. From these results it will be seen that while Q_{total}/Q_{Top} is always greater than approximately 4 it is only at low values of φ that the contribution of Q_{Top} to Q_{total} can be neglected. The assumption that the contribution of Q_{Top} to Q_{total} can be neglected has sometimes been used in deriving correlation equations for the heat transfer rate from vertical square cylinders using:

$$Nu = Nu_{side}\left(\frac{A_{side}}{A_{total}}\right) + Nu_{top}\left(\frac{A_{Top}}{A_{total}}\right) \tag{24}$$

and neglecting the second term on the right hand side. The present results show that this approach will not normally be applicable to the inclined cylinder case.

The surfaces that make up the square cylinder are of the form of flat plates connected to each other. The correlation equation for the case of a wide flat plate with a uniform surface temperature has the form:

$$Nu_0 = \mathrm{B}\, Ra^{0.25} \tag{25}$$

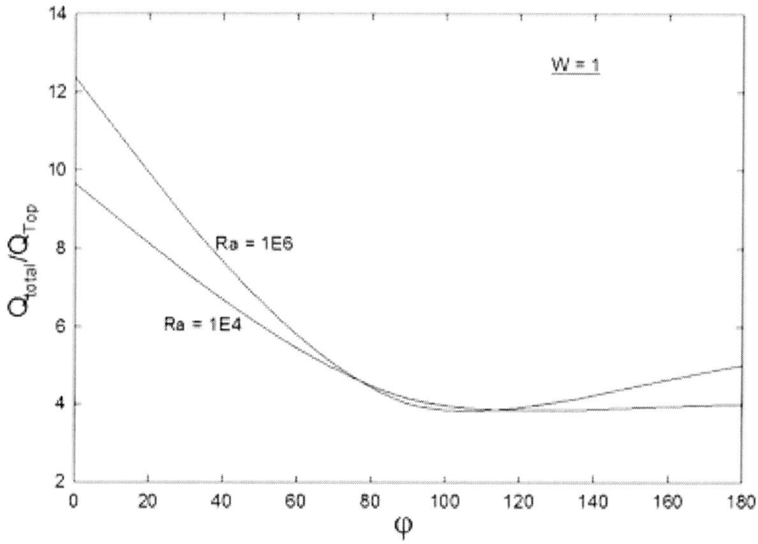

Figure 21. Variation of the ratio of the total heat transfer rate from the surface of the cylinder to the heat transfer rate from the *top* surface with angle of inclinations for *Ra* values of 10^4 and 10^6 and for $W = 1$

The parameter B depending only on the Prandtl number of the fluid involved. When the width of the plate becomes relatively small and inclined at an angle to the vertical edge effects become important (see Kalendar and Oosthuizen [46]) and the Nusselt number depends on the boundary layer thickness to plate width ratio, i.e., since the boundary layer thickness will be dependent on the value of:

$$\varepsilon = \frac{h}{wRa^{0.25}} = \frac{1}{WRa^{0.25}} \tag{26}$$

The Nusselt number will of course also be dependent on the inclination angle. Therefore when the surfaces that make up the square cylinder are narrow and inclined at an angle to the vertical, the heat transfer rate from the square cylinder is given by an equation of the form:

$$\frac{Nu}{Ra^{0.25}} = \text{constant} + \text{function}\ (Ra,\ W,\ \varphi) \tag{27}$$

Using an equation of this form it has been found that the present results for the case of a heated square cylinder with an exposed top surface and inclined at an angle, φ,of between $\varphi = 0°$ and $\varphi = 180°$, and with different dimensionless widths can be approximately described by:

$$\frac{Nu_{memp}}{Ra^{0.28}} = 0.27 + \frac{0.65}{(WRa^{0.25})^{0.95}} \tag{28}$$

The index on the left hand side of Eq. (28) has a Rayleigh number to the power of 0.28 instead of 0.25 since the interaction of flow over the plates that make up the square cylinder become more important as the plate width decreases. A comparison of the results given by this correlation Eq. (28) with the numerical and experimental results is shown in Fig. 22. It will be seen that the equation describes the computed numerical results to an accuracy of better than 92 %. Figure 23 shows the comparison of the experimental results with the correlation Eq. (28), the agreement being to an accuracy of better than 86 %, where the 12 % in Fig. 23 shows the 95 % prediction band of the experimental results.

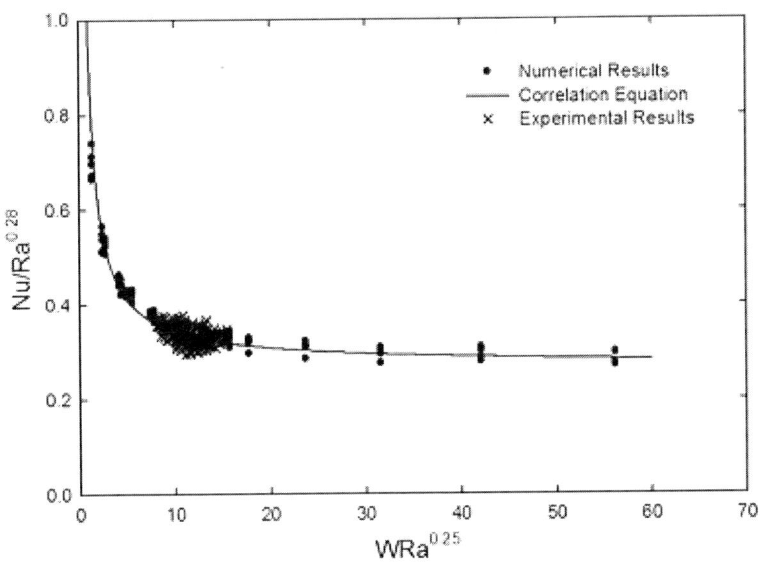

Figure 22. Comparison of correlation equation with the numerical and experimental for inclined square cylinder with exposed *top* surface

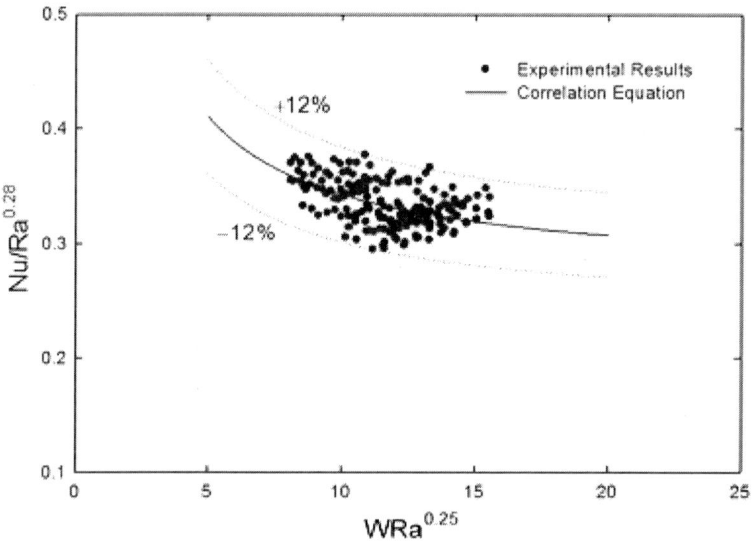

Figure 23. Comparison of correlation equation with the experimental results for inclined square cylinder with exposed*top* surface

5 CONCLUSIONS

Natural convective heat transfer from an isothermal inclined square cylinder which have an exposed top surface and is, in general, inclined at an angle to the vertical between vertically upwards and vertically downwards has been numerically and experimentally studied. The effects of Dimensionless widths, Rayleigh numbers, and inclination angles on the mean Nusselt number for the entire cylinder and for the mean Nusselt numbers for the various surfaces that make up the cylinder were conducted in this study. The results of the present study indicate that:

1. The experimental results are in a good agreement with the numerical results for square cylinder for all angle of inclination considered in this study.
2. The mean Nusselt number for the cylinder increases with decreasing W under all conditions considered.
3. The relative magnitudes of the mean Nusselt numbers for the various faces that make up the cylinder vary considerably with the inclination angle and as a result it is only possible to neglect the heat transfer from the 'top' surface compared to that from the

other surfaces when the dimensionless width is small and the angle of inclination is near 0°.

4. At lower values of *Ra* (approximately less than 10^4) for all dimensionless plate widths, *W*, considered the mean Nusselt number is independent of angle of inclination φ. At larger values of *Ra* the dependence of the mean Nusselt number on the angle of inclination φ becomes significant.
5. At the higher values of Rayleigh number the lowest mean Nusselt number occurs when the cylinder is in a horizontal position, i.e., when φ is equal 90° while the highest mean Nusselt number occurs when the cylinder is at inclination angles, φ, of 45° and 135°.
6. The results for the mean Nusselt number based on the mean heat transfer rate from vertical and inclined square cylinder can be adequately correlated by the Eq. (28).

ACKNOWLEDGMENTS

This work was supported by the Public Authority for Applied Education and Training of Kuwait and the Natural Sciences and Engineering Research Council of Canada.

REFERENCES

1. Ede AJ (1967) Advances in free convection. Adv Heat Transf 4:1–64
2. Burmeister LC (1983) Convective heat transfer. Wiley, New York
3. Jaluria Y (1980) Natural convection, 1st edn. Pergamon Press, Oxford
4. Cebeci T (1974) Laminar free convective heat transfer from the outer surface of a vertical slender circular cylinder. In: Proceedings of 5th international heat transfer conference, society of heat transfer of Japan, vol 3. Tokyo, Japan, Paper NC 1.4, pp 15–19
5. Lee HR, Chen TS, Armaly BF (1988) Natural convection along slender vertical cylinders with variable surface temperature. J Heat Transf Trans ASME 110(1):103–108

6. Le Fevre EJ, Ede AJ (1956) Laminar free convection from the outer surface of a vertical circular cylinder. In: Proceedings 9th international congress applied mechanics Brussels, vol 4. pp 175–183

7. Crane LJ (1976) Natural convection from a vertical cylinder at very large Prandtl numbers. J Eng Math 10(2):115–124

8. Kimura F, Tachibana T, Kitamura K, Hosokawa T (2004) Fluid flow and heat transfer of natural convection around heated vertical cylinders (effect of cylinder diameter). JSME Int J Ser B Fluids Therm Eng 47(2):156–161

9. Oosthuizen PH (2007) Natural convective heat transfer from an isothermal vertical cylinder with an exposed upper surface mounted on a flat adiabatic base. In: Proceedings ASME international mechanical engineering congress and exposition, IMECE, Seattle, Washington, USA, November 2007, Paper IMECE2007-42711, pp 389–395

10. Oosthuizen PH (1979) Free convective heat transfer from vertical cylinders with exposed ends. Trans Can Soc Mech Eng 5(4):231–234

11. Jarall S, Campo A (2005) Experimental study of natural convection from electrically heated vertical cylinders immersed in air. Exp Heat Transf 18(3):127–134

12. Welling I, Koskela H, Hautalampi T (1998) Experimental study of the natural-convection plume from a heated vertical cylinder. Exp Heat Transf 11(2):135–149

13. Fukusawa K, Iguchi M (1962) On optical measurements of natural convection along vertical cylinder. J Mech Lab Jpn 16(3):114–120

14. Oosthuizen PH (2007) Natural convective heat transfer from a vertical cylinder with an exposed upper surface. In: ASME/JSME thermal engineering summer heat transfer conference, ASME, Vancouver, BC, Canada, pp 489–495

15. Oosthuizen PH, Paul JT (2008) Natural convective heat transfer from an isothermal cylinder with an exposed upper surface mounted on a flat adiabatic base with a flat adiabatic surface above the cylinder. In: EUROTHERM, 5th European thermal science conference, Eindhoven, Netherlands

16. Koch W (1927) Heat transmission from hot pipes. Beih. Gesundh. Ingr. 22:1–27

17. Farber EA, Rennat HO (1957) Variation of heat transfer coefficient with length-inclined tubes in still air. Ind Eng Chem 49(3):437–440

18. Eckert ER (1959) Heat and mass transfer, 2nd edn. McGraw-Hill, New York

19. Oosthuizen PH (1976) Experimental study of free convective heat transfer from inclined cylinders. J Heat Transf 98(4):672

20. Al-Arabi M, Salman YK (1980) Laminar natural convection heat transfer from an inclined cylinder. Int J Heat Mass Transf 23(1):45–51

21. Al-Arabi M, Khamis M (1982) Natural convection heat transfer from inclined cylinders. Int J Heat Mass Transf 25(1):3–15

22. Chand J, Vir D (1980) A unified approach to natural convection heat transfer in the laminar region from horizontal and inclined cylinders. Lett Heat Mass Trans 7(3):213–225

23. Oosthuizen PH, Mansingh V (1983) Free and forced convective heat transfer from short cylinders. In: Joint ASME/AIChE national heat transfer conference ASME, Seattle, WA, USA, July 1983

24. Oosthuizen PH, Paul JT (1991) Natural convective heat transfer from inclined downward pointing cylinders with exposed ends. In: Proceedings of the world conference on experimental heat transfer, fluid mechanics, and thermodynamics. pp 697

25. Popiel CO, Wojtkowiak J (2004) Experiments on free convective heat transfer from side walls of a vertical square cylinder in air. Exp Therm Fluid Sci 29(1):1–8

26. Al-Arabi M, Sarhan A (1984) Natural convection heat transfer from square cylinders. Appl Sci Res 41(2):93–104

27. Oosthuizen PH (2008) Natural convective heat transfer from an isothermal vertical square cylinder mounted on a flat adiabatic base. In: ASME summer heat transfer conference, ASME, Jacksonville, FL, United states, August 2008, pp 499–505

28. Starner KE, McManus HN (1963) An experimental investigation of free convection heat transfer from rectangular fin arrays. J Heat Trans 85:273–278

29. Welling JR, Wooldridge CR (1965) Free convection heat transfer coefficients from rectangular vertical fins. J Heat Trans 87:439–444

30. Harahap F, McManus HN (1967) Natural convection heat transfer from horizontal rectangular fin array. J Heat Transfer 89:32–38

31. Leung CW, Probert SD, Shilton MJ (1986) Heat transfer performance of vertical rectangular fins protruding from bases (Effect of fin length). Appl Energy 22:313–318

32. Leung CW, Probert SD (1989) Thermal effectiveness of short-protrusion rectangular heat exchanger fins. Appl Energy 34:1–8

33. Yuncu H, Anbar G (1998) An experimental investigation on performance of rectangular fins on a horizontal base in free convection heat transfer. Heat Mass Transf 33:507–514

34. Mobedi M, Yuncu H (2003) A three dimensional numerical study on natural convection heat transfer from short horizontal rectangular fin array. Heat Mass Transf 39:267–275

35. Harahap F, Rudianto E (2005) Measurements of steady-state heat dissipation from miniaturized horizontally-based straight rectangular fin arrays. Heat Mass Transfer 41:280–288

36. Sahray D, Magril R, Dubovsky V, Ziskind G, Letan R (2006) Natural convection heat transfer from pin fin heat sinks with internal base. In: Proceedings of 13th international heat transfer conference, Sydney, Australia

37. Harahap F, Lesmana H, Dirgayasa S (2006) Measurements of heat dissipation from miniaturized vertical rectangular fin arrays. Heat Mass Transf 42:1025–1036

38. Yazicioglu B, Yancu H (2007) Optimum fin spacing of rectangular fins on a vertical base in free convection heat transfer. J Heat Mass Transf 44:11–21

39. Sahray D, Magril R, Dubovsky V, Ziskind G, Letan R (2007) Study of horizontal-base pin-fin heat sinks in natural convection. In: Proceedings of ASME InterPACK, Vancouver, British Columbia, Canada, July 2007, pp 925–931

40. Kalendar A, Oosthuizen PH (2009) Natural convective heat transfer from an inclined isothermal square cylinder with an exposed top surface mounted on a flat adiabatic base. In: Proceedings of the ASME heat transfer summer conference, San Francisco, California USA, July 2009, HT2009-88094, pp 115–122

41. Kalendar A, Oosthuizen PH, Alhadhrami A (2010) Experimental study of natural convective heat transfer from an inclined isothermal square cylinder with an exposed top surface mounted on a flat adiabatic base. In: Proceeding of the International heat transfer conference IHTC14, Washington, DC, USA, August 2010, Paper IHTC14-22846, pp 113–120

42. Kalendar A (2011) Numerical and experimental studies of natural convective heat transfer from vertical and inclined flat plates and short cylinders. PHD thesis, Queen's University, Kingston, Canada

43. Moffat RJ (1983) Using uncertainty analysis in the planning of an experiment. Trans ASME J Fluids Eng 107(2):173–178

44. Moffat RJ (1988) Describing the uncertainties in experimental results. Elsevier Sci Publ Exp Therm Fluid Sci 1:3–17

45. Churchill SW, Chu HHS (1975) Correlating equations for laminar and turbulent free convection from a vertical plate. Int J Heat Mass Transf 18:1323–1329

46. Kalendar A, Oosthuizen PH (2011) Numerical and experimental studies for natural convective heat transfer from vertical and inclined narrow isothermal flat plates. Heat Mass Transfer 47:1181–1195

8

CHAPTER

3-D Numerical Simulation of Heat Transfer in Biomedical Applications

Aleksandra Rashkovska, Roman Trobec, Matjaž Depolli and Gregor Kosec

[1] *Jožef Stefan Institute, Slovenia*

1. INTRODUCTION

Computational modelling has received significant attention in the research community over the past few decades due to its pronounced contribution in better understanding of the nature, as well as in the development of advanced technologies. The modelling of more and more complex transport physical systems helps the community to address important issues, like identification of environmental problems, improvement of technological processes, development of biomedical applications, etc. However, the physical modelling is only one part of the entire problem. In most realistic scientific computing applications, a physical model cannot be solved in a closed form, thus adequate numerical approach is required. The continuous physical domain is replaced by its discrete form. In the majority of the simulations, the classical numerical methods, such as the Finite Volume Method (FVM) (Ferziger and Perić, 2002), Finite Difference Method (FDM) (Őzisik, 2000, Shashkov, 1996), Boundary Element Method (BEM) (Białecki, *et al.*, 2006, Wrobel, 2002) or the Finite Element Method (FEM) (Rappaz, *et al.*, 2003), are used to solve these problems.

An important part of the numerical approach is the effective implementation of the solution procedure on modern computer architectures. Realistic computations might take vast amount of time to complete, especially in 3-D domains with high number of computational nodes. The computing performance can be improved by numerous approaches, such as network computing, grid computing, cloud computing and other variants of "distributed computing" (Trobec, *et al.*, 2009).

In recent decades, computer simulations have proved a great help in understanding and solving a variety of problems in science. Advances in computer technology enable simulation of natural phenomena that cannot be subject to experiment in reality since experiments would be ecologically problematic, dangerous to humans, or would cost vast amounts of money (Martino, *et al.*, 1994). Especially in medicine, experiments are often difficult to perform because human subjects are involved. Measurements during clinical procedures are time consuming and often not as accurate as desired, because many parameters are difficult to control (McMaster, *et al.*, 1978). Many examples can be found in medicine, where in vivo experiments and measurements are often difficult, dangerous or even impossible (Trunk, *et al.*, 2003), especially if deep tissues or vital organs are in question (Tikuisis and Ducharme, 1991). However, simulation can provide the only safe and inexpensive insight into physiological processes. With the use of computer simulation, it is possible to calculate, analyse, and visualize both stationary temperature fields and the changes of temperature in time in living biological tissues (Grana, 1994). The temperature of the human tissue is an important factor in many fields of physiology, sports (Olin and Huljebrant, 1993), cryotherapy (Trobec, *et al.*, 1998), etc. It has been recognized that the temperature field is influenced by the environmental conditions, the temperatures of neighbouring tissues, the muscle metabolism and the blood circulation. Different tissues have different physical and thermodynamic properties and respond diversely to temperature changes (Bernardi, *et al.*, 2003, Pennes, 1948, Trunk, *et al.*, 2003). The temperature field varies in space and time in different parts of the investigated domain.

In this chapter we present numerical solution procedure for the treatment of heat transfer in biological tissues. Different body parts are modelled and given as numerical examples. All cases are focused on realistic three dimensional (3-D) models and realistic process parameters. In all numerical examples we use 3-D closed cavity as a computational domain. Within the domain, high resolution models of body parts are used. The models are obtained through expert assisted segmentation of human body parts from the Visible Human Dataset (VHD) (Ackerman, 1998). The body parts are composed of different tissues (muscles, bones, cartilage, fat, vessels and nerves). To respect the complexity of the modelled body parts, a one-domain approach is employed through spatial dependent thermo-physical properties. The final geometrical domain is composed of small cubic voxels, resulting in millions of discretization points.

The Pennes' Bio-heat equation (Pennes, 1948) represents the basis for the physical model. It describes the energy continuity within biological tissues by incorporating heat convection, heat transfer between blood and tissues, and heat production by metabolism. Some important extensions have been introduced, in particular, an inhomogeneous spatial model composed of tissues with different characteristics and modelling of the heat contributions from arteries, blood perfusion, and metabolism as functions of the surrounding tissue temperature. This improved Bio-heat equation was evaluated in terms of sensitivity and accuracy and solved numerically on single and parallel computers (Akl, 1997,Martino, *et al.*, 1994). Outside biological tissues, the Bio-heat equation is reduced to diffusion equation as both additional source terms (heat transfer between blood and tissues, and metabolism) vanish. The flow outside tissues is modelled as the Navier Stokes fluid (Ferziger and Perić, 2002), while in the tissue no flow is allowed. We consider only incompressible fluids.

A substantial amount of work on closed form and numerical solutions of the Bio-heat equation has been published (Bernardi, *et al.*, 2003, Bowman, *et al.*, 1975, Charny, 1992, Pennes, 1948). In our examples, the governing system of partial differential equations (PDEs) is numerically solved through spatial discretization with the FDM and the Euler time stepping (Őzisik, 2000). We present three different numerical studies regarding different topics in medicine, where results are computed using the proposed physical model and numerical solution procedure.

The first numerical example is dedicated to a study of the temperature changes in a resting proximal human forearm. By computer simulation, we explained the variations in the Pennes' well-known in vivo measurements of the steady-state temperatures along the transverse axis of the proximal forearm (Pennes, 1948). Such in vivo measurements are no longer practiced and are very valuable, as are their additional analyses. The domain of the simulation is a region of the forearm 5 cm – 15 cm below the elbow. We use Robin boundary conditions on the contact between the skin and the ambient air, while on the domain boundaries the heat flow is set to zero. The heat production by tissue metabolism is modelled using the Q10 rule (Tikuisis and Ducharme, 1991), while the heat exchange between the blood and the tissue is modelled as a function of the local temperature and the regional blood flow. We confirm the stability and accuracy of the method by varying the simulation parameters, the initial and boundary values, and the model dimensions, with subsequent analysis of the results. Our simulations indicate that the fluctuations of the measured steady-state temperatures are a natural consequence of a complex interplay between the position of the measuring probes, the anatomical position of the main arteries, the dimensions of the forearm, the blood flow, the inhomogeneity of tissues, and the environmental temperature.

The second illustrative example is simulation of a human knee exposed to topical cooling after anterior cruciate ligament (ACL) surgery (Martin, *et al.*, 2001, McMaster, *et al.*, 1978, Tikuisis and Ducharme, 1991, Trobec, *et al.*, 2008). The computational domain is composed of the knee model and surrounding layers: protective bandage, cooling layer, isolating blanket and ambient air. The ambient air is held at constant initial temperature as it is assumed to be well mixed. The heat flux from the first and last slices is kept constant in order to imitate the influence of the leg not exposed to the cooling. Simulations are performed on the described model using realistic process parameters gathered from measurements. Two different methods of topical cooling are compared; first, the use of a gel-pack filled with refrigerated gel, which is exposed to ambient temperature and therefore becomes less and less effective as the gel is warming; and second, the use of a cryo-cuff cooled by a liquid at constant temperature maintained by an external cooling device. The results are presented in terms of temporal temperature development in critical points of the knee. The temperature dynamics is also validated by in vivo temperature measurements during cryotherapy for different regimes of cooling. We show that the second cooling approach gives much better results regarding the desired temperature distribution within the knee.

The cooling of a human heart during surgery is taken as a final, most complex example. In the last example full model is incorporated. Besides the Bio-heat equation Navier Stokes fluid flow is solved as well. The topical cooling (Olin and Huljebrant, 1993) is simulated through consideration of partially submerged heart in the cooling liquid. The effect of natural convection in the cooling liquid is tackled. We show that considering solely diffusive heat transport, as done in present (Šterk and Trobec, 2005,Trobec, *et al.*, 1998, Trunk, *et al.*, 2003), simulation does not supply adequate results.

The presentation of the numerical examples is organized as follows. First, a short historical overview of the tackled subject is introduced. Next, the spatial 3-D geometrical model of the designated body part is described. Then, different simulation modalities and parameters are stated. Afterwards, analysis of the simulated results and their comparison with the measured values, if available, is presented. Finally, conclusions are presented.

Before presenting the numerical examples, the considered physical model and solution procedure are introduced.

2. PHYSICAL MODEL

The present chapter is focused on the formulation of the physical model suitable for describing biological tissues. The main part of the modelling is heat continuity, where biological response is treated as volumetric heat source. The power of such sources depends on the temperature of the tissue. Besides bio-heat transfer within tissues, the surrounding is modelled as a fluid in order to simulate the heat exchange with the ambient. The surrounding is treated as a Newtonian fluid described by Navier-Stokes equation. The fluid is assumed to be incompressible.

The model in the present chapter is developed from the classical continuum conservation equations for energy, momentum and mass. The differential forms of the conservation equations are:

$$\frac{\partial}{\partial t}\left(\rho c_p T\right)+\nabla\cdot\left(\rho c_p \mathbf{v}T\right)=-\nabla\cdot\mathbf{j}+S, \tag{1}$$

$$\frac{\partial}{\partial t}\left(\rho \mathbf{v}\right)+\nabla\cdot\left(\rho \mathbf{v}\mathbf{v}\right)=-\nabla P+\nabla\cdot\boldsymbol{\tau}+\mathbf{b}, \tag{2}$$

$$\frac{\partial \rho}{\partial t}+\nabla\cdot\left(\rho \mathbf{v}\right)=0, \tag{3}$$

for energy (1), momentum (2) and mass (3), where ρ,t,v,P,τ,b,S,cp,Tandjρ,t,v,P,τ,b,S,cp,Tandj stand for density, time, velocity, pressure, deviatory stress tensor, body force, volumetric heat sources, specific heat, temperature and heat flux, respectively.

2.1. Model Assumptions

To construct an accurate and stable model, on the one hand, and still solvable in a reasonable way, on the other hand, some basic assumptions are made:

- all material properties are assumed to be temperature independent,
- no flow is allowed in the tissues,
- Newtonian fluid is considered,
- the biological response of the tissues on temperature change is modelled with the Pennes' Bio-heat equation,
- the metabolic heat production per unit mass hmhm is assumed to obey the Q10 rule,
- all materials are isotropic.

2.2. Bio-Heat Transport

The local mass heat flux is modelled by the Fourier law

$$\mathbf{j} = -\boldsymbol{\lambda}\nabla T\,, \tag{4}$$

with λλ standing for the thermal conductivity tensor. Considering the model assumption for material isotropy, the thermal conductivity diffusion tensor is rewritten in the following form:

$$\boldsymbol{\lambda} = \lambda \mathbf{I}\,, \tag{5}$$

where IandλIandλ stand for identity matrix and thermal conductivity.

The biological response to the conditions within the tissue is modelled through the volumetric sources introduced by Pennes (Pennes, 1948, Wissler, 1998):

$$S = \underbrace{v_r \frac{\rho^b c_p^b}{\rho c_p} V_r(T)(T_a - T)}_{\textit{blood perfusion}} + \underbrace{h_r \frac{1}{c_p} h_m(T)}_{\textit{metabolic heat production}}\,, \tag{6}$$

where ρbρb and cbpcpb are the density and the specific heat of the blood, and SS stands for the heat generated from the biological response. The dimensionless parameters hrhr and vrvr are used to control the spatial/material dependency of the biological response and were not included in the original paper. The functions VrandhmVrandhmcharacterize the behaviour of heat generation with respect to the temperature of the tissue and are defined latter in this chapter. The temperature Ta=36.8∘CTa=36.8∘C stands for the reference tissue temperature.

Combining the equations (1),(4),(5) and (6) yields the well-known Pennes Bio-heat equation (Pennes, 1948, Wissler, 1998):

$$\frac{\partial T}{\partial t}=\underbrace{\frac{1}{\rho c_p}\nabla\cdot\left(\lambda\nabla T\right)-\nabla\cdot\left(\mathbf{v}T\right)}_{heat\ transport}+\underbrace{v_r\frac{\rho^b c_p^b}{\rho c_p}V_r(T)(T_a-T)+h_r\frac{1}{c_p}h_m(T)}_{\text{biological response}}. \tag{7}$$

Equation (7) models the heat transfer between the arterial blood and the tissues, and the heat production arising from tissue metabolism, in addition to the classical heat transport.

To close the energy transport model, both biological response functions have to be defined.

Published measurements of the blood flow rate (Vr)(Vr) show that it increases with the tissue temperature (Ducharme and Frim, 1988). Earlier measurements of the forearm blood flow were limited to plethysmography that measures the mean total blood flow through a given region of the forearm (Barcroft and Edholm, 1942). Later, more data became available also for regional blood flow of specific tissues (Cooper, *et al.*, 1955, McElfresh and Kelly, 1974, Sugimoto and Monafo, 1987). Several studies have established that the muscle blood flow does not increase with temperature; most of the increase takes place in the skin (Detry, *et al.*, 1972). The value of the mean blood flow rate V as a function of the skin temperature T_s has been reported in several publications (Barcroft and Edholm, 1946, Cooper, *et al.*, 1955, Ducharme and Tikuisis, 1994, Wenger, *et al.*, 1986, Yue, *et al.*, 2007). We use approximate exponential function:

$$V\left(T\right)=\left(V_1 e^{V_2 T}+V_3\right)V_0, \tag{8}$$

with newly introduced case dependant dimensionless parametersV1,V2and V3V1,V2andV3. The reference value is set to V_0 = 1.667 10^{-4} s^{-1}. Although a more detailed blood flow model could be used in our simulation, for example, incorporating the autonomic thermoregulation of the body, its development is beyond the scope of this chapter.

The metabolic heat production in the human body can be divided into unregulated heat production from voluntary muscle contraction and normal metabolic pathways, and into regulated heat production for maintaining temperature homeostasis at lower ambient temperatures (Bullock and Rosendahl, 1992,Proulx, *et al.*, 2003). The rate of metabolic heat production per unit mass hmhm is assumed to obey the Q10 rule (Bullock and Rosendahl, 1992), which is expressed as a function of the tissue temperature:

$$h_m(T) = h_0 2^{\frac{T-T_r}{10}}$$

(9)

where h0=1W/kgh0=1W/kg stands for the reference metabolic heat production of a tissue at reference temperature Tr = 35 °C. An example of the biological response against temperature change is presented in Figure 1, where the parameters are taken as in the following first numerical example.

Outside the tissue, the Bio-heat equation is reduced to the classical heat transfer equation (Amimul, 2011), as both biological sources vanish:

$$\frac{\partial T}{\partial t} = \frac{1}{\rho c_p} \nabla \cdot (\lambda \nabla T) - \nabla \cdot (\mathbf{v} T) \ .$$

(10)

The model assumes no flow in the tissues and, as a consequence, the advection term vanishes. In future discussions, a tissue is referred as a solid material.

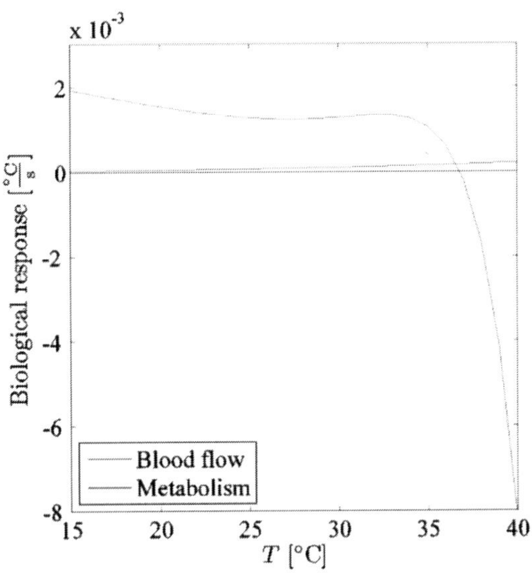

Figure 1. The modelled biological response composed of the blood perfusion and the metabolism as a function of the muscle temperature.

2.3. Fluid Flow

In order to specify the momentum equation (2), the stress tensor has to be specified. It is constituted as a Newtonian fluid:

$$\tau_{ij} = \mu\left(\frac{\partial v_i}{\partial p_j} + \frac{\partial v_j}{\partial p_i}\right), \tag{11}$$

where μμ stands for the dynamic viscosity. Further, the body force in the momentum conservation equation is modelled by the Boussinesq approximation for temperature dependent density:

$$\mathbf{b} = \mathbf{g}\rho_{ref}\left[1 + \beta_T\left(T - T_{ref}\right)\right], \tag{12}$$

where ρrefandTrefρrefandTref stand for the reference values and gg for the gravitational acceleration. Combining the equations (2), (11) and (12) yields:

$$\frac{\partial}{\partial t}(\rho\mathbf{v}) + \nabla\cdot(\rho\mathbf{v}\mathbf{v}) = -\nabla P + \nabla\cdot(\mu\nabla\mathbf{v}) + \mathbf{g}\rho_{ref}\left[1 + \beta_T\left(T - T_{ref}\right)\right]. \tag{13}$$

To close the system, the mass continuity (3) is considered. The model assumes incompressible fluids, thus the velocity is divergent free:

$$\nabla\cdot\mathbf{v} = 0\,. \tag{14}$$

2.4. Thermo-Physical Properties

The thermo-physical properties of living tissues are still not uniformly defined. For present simulation purposes, the average of published data is used (Bowman, *et al.*, 1975, Moses and Geddes, 1990, Ponder, 1962, Poppendiek, *et al.*, 1967)]. For muscles at rest, h_r was taken to be about half the human basal metabolic rate, which equals 0.6 W/kg (Harris and Benedict, 1918, Ho, *et al.*, 1994, Silverthorn, 2001). The impact of the metabolic rate production is almost negligible for the temperature field in the forearm; even so, it can be adjusted in the model for each tissue independently.

Table 1. Thermo-physical properties

	$\lambda\left[\frac{W}{mK}\right]$	$c_p\left[\frac{Jkg}{K}\right]$	$\rho\left[\frac{kg}{m^3}\right]$	v_r	h_r
Skin	0.510	3431	1200	0.4	0.3
Subcutaneous/adipose tissue	0.550	2241	812	0.4	0.3
Muscle	1.030	4668	1179	0.8	0.6
Cortical bone	2.280	1260	1700	0.2	0.1
Cancellous bone	0.500	2260	900	0.6	0.3
Nerve	0.500	3277	1190	1.1	0.3
Blood	0.670	3890	1057	/	/
Ambient air	0.025	1012	1.29	/	/
Water	0.580	4204	1000	/	/

3. SOLUTION PROCEDURE

3.1. Geometric Model Preparation

The computer model domain of the forearm, the knee, and the heart is derived from the axial anatomic images of a male human body that are available from the Visible Human Dataset (VHD) (Ackerman, 1998). A similar model of the whole body at a resolution of 1 mm^3 has been developed within the VHD project, but was not available for public use at full extent when this chapter was written. The axial anatomic images are 24-bit colour cross-sections photographs of a frozen male human cadaver, with resolution of 0.33 mm ×0.33 mm, and taken at spatial interval of 1 mm.

To construct the three-dimensional model, several consecutive slices (cross-sections) (Figure 2) of VHD are used. First, exact overlapping of all slices is required to ensure the integrity of the model. Therefore, a reference point on every slice is needed to align the slices. The white circle in the top right-hand corner of the VHD cross-section is selected for this purpose. A simple automatic procedure is used to find the centre of this circle and thus the absolute position of each slice. The advanced elastic matching (Dann, *et al.*, 1989) is not applied at this stage. Second, several impurities and other errors in the original photographs have to be discarded manually. Last, different tissues have to be identified on the slices. This is done manually, under the supervision of an expert for anatomy, with image editing software, by painting over the photographs with predefined colours, always the same colour for the same tissue. Only the most prominent tissues are marked, the rest are approximated with the surrounding tissue. An automated procedure is then applied to scale down the slices to 1 mm×1 mm resolution, convert them from colour images to a format recognized by the computer simulator, and stack them together into a 3-D model. Finally, the 3-D model is checked and edited in custom modelling software to smooth out the jaggedness in *z* axis appearing as a result of the previous processing done only on *x*-*y* cross-sections.

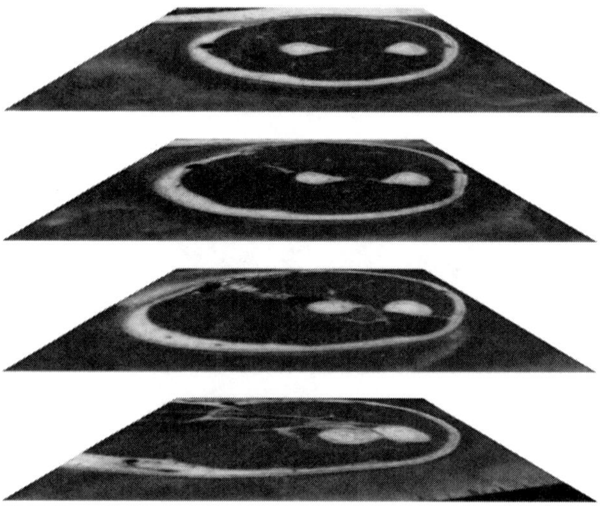

Figure 2. Example of stacked images of VHD for arm.

Each voxel of the geometric model is characterized by thermo-physical properties.

3.2. Discretization

The described physical model does not have a closed form solution and has to be treated by a numerical approach. The present work deals with the transient problems described by parabolic partial differential equations.

Such problems are space and time dependent and therefore discretization in space and time is required.

In the present solution procedure, one of the simplest and most intuitive numerical techniques for time dependent simulation is used. Temporal discretization is performed through two-level explicit discretization. The most commonly and widely used unconditionally stable schemes for temporal stepping are implicit and Crank-Nicolson approaches. Besides two-level methods, also multilevel methods can be used to achieve higher-order approximations. The disadvantage of the multilevel methods is the need to store data from several previous time steps. The multilevel data storage issue can be avoided by using Runge-Kutta methods.

For the spatial discretization, the Finite Difference Method (FDM) (Őzisik, 2000, Shashkov, 1996) is used, as it has low calculation complexity, is sufficiently accurate, and is simple to implement. In addition, FDM can directly use the presented discretized domain, where uniform discretization elements are used. There are also several numerical techniques with higher order of accuracy and better stability properties. For example, spatial discretization can be performed through weak form of Finite Element Method (FEM) (Rappaz, *et al.*, 2003) or Boundary Element Method (BEM) (Białecki, *et al.*, 2006, Wrobel, 2002). Furthermore, there are several meshless techniques (Atluri and Shen, 2002, Atluri, 2004, Chen, 2002, Kansa, 1990, Liu, 2003, Šarler, 2007) with convenient properties.

However, the complication in the temporal and/or spatial discretization of the domain gains benefits in accuracy and stability, but loses much in terms of computational time. In the light of this work, the simplest two-level method is used with the lowest CPU complexity.

In general, a partial differential equation

$$\frac{\partial \Theta(\mathbf{p},t)}{\partial t} = L\left(\Theta(\mathbf{p},t)\right) \qquad (15)$$

is solved. In equation (15) Θ,Landp(x,y,z)Θ,Landp(x,y,z)stand for an arbitrary field/quantity, a second order partial differential operator and a position vector, respectively. In the present work, a transport equation is treated, which is governed by the Laplace(∇2)(∇2), the gradient (∇)(∇) and the divergence(∇·)(∇·) operators. First and second spatial derivatives are needed to construct those operators. Both derivatives are evaluated through Taylor's expansion, known as finite differences:

$$\frac{\partial}{\partial p_{\varepsilon}}\Theta(\mathbf{p})=\frac{\Theta(\mathbf{p}+\Delta\mathbf{p}_{\varepsilon})-\Theta(\mathbf{p}-\Delta\mathbf{p}_{\varepsilon})}{2\Delta p_{\varepsilon}}, \tag{16}$$

$$\frac{\partial^2}{\partial^2 p_{\varepsilon}}\Theta(\mathbf{p})=\frac{\Theta(\mathbf{p}+\Delta\mathbf{p}_{\varepsilon})-2\Theta(\mathbf{p})+\Theta(\mathbf{p}_{\varepsilon-1}-\Delta\mathbf{p}_{\varepsilon})}{\Delta {p_{\varepsilon}}^{2}}. \tag{17}$$

In equations (16) and (17), the index εε denotes spatial direction(x,y,z)(x,y,z),pεpε stands for the εε component of the position vector,ΔpεΔpε stands for the spatial step in εε direction, and ΔpεΔpε stands for the offset in εεdirection.

The left side of equation (15) is discretised through two-level time discretization with the finite difference representation of the first temporal derivative. It is rewritten in an explicit form as:

$$\frac{\Theta(\mathbf{p},t+\Delta t)-\Theta(\mathbf{p},t)}{\Delta t}=L\big(\Theta(\mathbf{p},t)\big)+O(\Delta t). \tag{18}$$

For the finite difference spatial discretization, the Neumann (or Fourier) stability analysis shows that both implicit and Crank-Nicolson are unconditionally stable, while the explicit scheme suffers from stability problems (Smith, 2003). For diffusive-convective problems, as the ones considered in the present work, the differential operator LL is generally written as:

$$L=\nabla\cdot\left(\frac{\lambda}{\rho c_p}\nabla(\ldots)\right)-\nabla\cdot(\rho\mathbf{v}\ldots), \tag{19}$$

and for the FDM, the stability criteria derived by Fourier analysis is stated as:

$$\frac{\lambda}{\rho c_p}\frac{\Delta t}{\Delta p_{\varepsilon}^{2}}\leq\frac{1}{6} \qquad |v_{\varepsilon}|\frac{\Delta t}{\Delta p_{\varepsilon}}\leq 1, \tag{20}$$

wherevεvεstands for the εε component of the velocity.

Despite the stability issues, in the present work the explicit method is used because of its straightforward local applicability and simplicity. The main drawback of the implicit and semi-implicit forms is the need for solving the global problem, where systems (usually large) of linear equations have to be considered in order to proceed to the next simulation time-step. On the other hand, the explicit scheme uses series of pre-solved linear systems solutions. In other words, the explicit scheme uses the local information from the current time-step (t0t0) to compute spatial derivatives, while the implicit scheme uses the current local time-step information only for the evaluation of time derivative. Finally, the Crank-Nicholson scheme uses linear combination of the current and the next time-step to improve the accuracy of the method.

One of the most convenient properties of the Explicit FDM (EFDM) is its straightforward parallel implementation, as there are no requirements of any kind for global communication.

3.3. Pressure Velocity Coupling

An important part of the solution procedure represents the solution of the momentum equation. An additional problem lies in the momentum-mass conservation coupling. The pressure is not explicitly included in the continuity equation and therefore, a special treatment is needed to solve the pressure-velocity coupling problem. The most widely known solution for the problem is the Semi-Implicit Method for Pressure Linked Equations SIMPLE (Ferziger and Perić, 2002), where the problem is translated to the Poisson equation problem. However, the SIMPLE algorithm demands the solution of the global system, which is complex from parallel implementation point of view. There are local alternatives, like the Local Pressure Velocity Coupling (LVPC) used in context with local meshless methods (Kosec and Šarler, 2008a), the artificial compressibility method (ACM), which has been recently under intense research (Malan and Lewis, 2011, Rahman and Siikonen, 2008, Traivivatana, *et al.*, 2007), and the framework for the Finite Difference Method in the SOLA approach (Hong, 2004). In this chapter, the more standard and well accepted pressure velocity coupling, based on the SIMPLE algorithm, is used.

The problem domain is discretized to voxels that can be either liquid or solid. For solid voxels only the velocity is assumed to be zero and no pressure computations take place. Otherwise, standard staggered grid approach (Ferziger and Perić, 2002, Shashkov, 1996) is used, with the temperature and the pressure defined in the middle of each voxel, and the velocity components defined on the respective perpendicular voxel edges.

The Navier-Stokes equation is discretized in time by using the explicit time discretization:

$$\frac{\hat{\mathbf{v}}-\mathbf{v}^{0}}{\Delta t}=\frac{1}{\rho}\left(-\nabla P^{0}+\nabla\cdot\left(\mu\nabla\mathbf{v}^{0}\right)+\mathbf{b}^{0}-\nabla\cdot\left(\rho\mathbf{v}^{0}\mathbf{v}^{0}\right)\right), \tag{21}$$

where v^ stands for the intermediate velocity and the superscript 0 denotes that the current time-step values are used. As the explicit time discretization is employed, the new time-step velocity is computed from the velocity values in the previous time-step. However, instead of a new time-step value, the intermediate velocity is introduced, which does not take into account the mass continuity (3). This problem is solved by introducing the velocity and pressure corrections. The added velocity correction drives the intermediate velocity towards divergent free field:

$$\nabla\cdot\left(\hat{\mathbf{v}}+\widehat{\mathbf{v}}\right)=0, \tag{22}$$

$$P^{1}=P^{0}+\widehat{P}, \tag{23}$$

where v⌢v⌢ and Unexpected text node: 'P 'Unexpected text node: 'P ' stand for the velocity and the pressure correction terms, and the superscript 1 denotes the next time-step. With the corrected velocity and pressure, the momentum equation can be rewritten as:

$$\underbrace{\frac{\hat{\mathbf{v}}-\mathbf{v}^{0}}{\Delta t}}_{I}+\underbrace{\frac{\widehat{\mathbf{v}}}{\Delta t}}_{II}=\underbrace{\frac{1}{\rho}\left(-\nabla P^{0}+\nabla\cdot\left(\mu\nabla\mathbf{v}^{0}\right)+\mathbf{b}^{0}-\nabla\cdot\left(\rho\mathbf{v}^{0}\mathbf{v}^{0}\right)\right)}_{I}\underbrace{-\frac{1}{\rho}\nabla\widehat{P}}_{II}. \tag{24}$$

$$\underbrace{\widehat{\mathbf{v}}}_{II}=\underbrace{-\frac{\Delta t}{\rho}\nabla\widehat{P}}_{II}, \tag{25}$$

thus all other terms subtract to zero. By applying the divergence over equation (25) and taking into account equation (22), the pressure correction equation emerges to:

$$\nabla\hat{\mathbf{v}} = \frac{\Delta t}{\rho}\nabla^2\bar{P} \,. \tag{26}$$

In order to solve the Poisson equation, additional boundary conditions are required. The boundary conditions for pressure field can be formulated by multiplying the momentum equation with the normal boundary vector. The boundary conditions are obtained with a similar approach used to construct the equation for pressure correction, stated as:

$$\frac{\partial\bar{P}}{\partial\mathbf{n}} = \frac{\rho}{\Delta t}\left(\hat{\mathbf{v}} - \mathbf{v}_{\Gamma}\right)\cdot\mathbf{n} \,, \tag{27}$$

where vΓvΓ stands for the velocity boundary condition and nn denotes the normal boundary vector. It is common to use the normal pressure boundary condition when no-slip and no-permeable velocity boundary conditions (vΓ=0) are used:

$$\frac{\partial}{\partial\mathbf{n}}\bar{P} = 0 \,. \tag{28}$$

With the pressure corrected according to (23), the velocity can be corrected as:

$$\mathbf{v}^1 = \hat{\mathbf{v}} - \frac{\Delta t}{\rho}\nabla\bar{P} \,, \tag{29}$$

where v1v1 stands for the next time-step velocity. With computed velocity and pressure, the simulation can proceed to the next time-step. The computation cycle is closed.

3.4. Implementation Of Parallel Program

The simulator is written in C++ and compiled with GCC. The parallelization is done by domain decomposition. The domain decomposition is one-dimensional: the domain is split in the *z* axis into subdomains of approximately equal size. The number of subdomains equals the number of processors participating in the solver execution. The sizes of the subdomains are determined at the program initialization and are constant for the whole execution; therefore, the load-balancing of the processors is static. One-dimensional decomposition was chosen because of its simplicity. To calculate a point on the border of a subdomain, data from one or more points from neighbouring subdomains might be required. In such cases, point-to-point communication between processors of those neighbouring subdomains is performed. MPI library is used for communication. It enables the program to execute in parallel on multicore computers, computer clusters or other distributed computer systems.

The multigrid method, used in our solver, is parallelized using similar principles. Calculations on fine grids are done in parallel using the same domain decomposition, while a single processor does the computation on the coarsest grid, which is negligible compared to the computation on the finest grid.

The developed physical model and solution procedure have been evaluated through comparing some of the simulated results against the measurements. The well-known Pennes (Pennes, 1948) in vivo measurements have been reproduced with the proposed model. The model results have been also compared with real case measurements of temperature distribution within the knee after the ACL surgery. The steady state profiles and the temporal transients have been analysed. For all cases, it is shown that an appropriate modelling of the bio-heat energy transport results in a good agreement with experimental data. The fluid flow solution procedure, however, is designed as a proven approach and tested on several standard benchmark tests, e.g., driven cavity, de Wahl Davis test, etc. The presentation of benchmark tests are beyond the scope of this chapter and are as such omitted. All results related to the bio-heat transfer are presented and elaborated in details in the following sections.

The numerical scheme used to solve the proposed physical model is already well accepted and extensively tested in the past. A detailed convergence of several numerical techniques applied the diffusion problem was published in (Trobec, *et al.*, 2012). It was shown that the FDM has an adequate convergence rate. Based on the obtained results and taking into account that we use the same approach we do not insert the convergence analysis in this chapter.

4. HEAT TRANSFER IN RESTING HUMAN FOREARM

The invaluable measurements of the steady-state temperature fields in forearms of subjects that are not anesthetized, published sixty years ago by Pennes (Pennes, 1948), have still not been explored in all details (Wissler, 1998). Pennes' used a thermocouple probe on a wire that was pulled through the forearm in a straight line (referred to as the measurement path in the rest of the section) in steps of less than 1 cm. At each step, the measurement was read after the probe temperature has stabilized. InFigure 7 the original measurements and the author's interpolated temperature curves are reproduced.

Other experimental data are available for the temperature of the forearm during an immersion in water at various temperatures, either as it evolves in time (Barcroft and Edholm, 1946) or near its steady-state. Since then, these measurements have been elaborated by others (Ducharme, 1988). In some recent studies, measured lower resolution temperature profiles in the forearm cross-sections are reported, together with estimated values for the in-vivo thermal conductivities of muscle and subcutaneous tissues (Ducharme, 1988). We recreated the conditions similar to those of the Pennes' measurements and simulated the steady-state temperatures of the proximal forearm. In this section, we consider the effects of blood perfusion, metabolism, position of the arteries, and inhomogeneity of tissues on the steady-state temperature profile of the forearm.

4.1. Geometric Model Of The Forearm

Cross-sections of the VHD male thorax (numbered 1601 – 1800) are used for the forearm model. The right forearm is cropped from those cross-sections as rectangles of 506 (width) ×632 (height) pixels. The VHD right forearm is inclined by 32.5° in the *x-z* plane and by 43.1° in the y-*z* plane. It is more suitable for the simulation if the *z* axis of the forearm is parallel to the *z* axis of the model; therefore, the elements of the model are sampled out of the forearm rectangles at an angle, making the *z* axes parallel. Finally, a regular grid of size 140×116×100 elements with spatial step of 1 mm is created. For the 2-D simulations, "model slices" are used – individual *x*-*y* planes of the grid, numbered by their position on the *z* axis.

The 3-D domain represents a section of the human forearm, 50 mm distal to the tip of the elbow and 100 mm long (50 mm from the probe insertion point in both directions). For better visibility of the bones and arteries in the model, muscles, skin, and subcutaneous tissues are shown semi-transparent. InFigure 3 model example slice is shown.

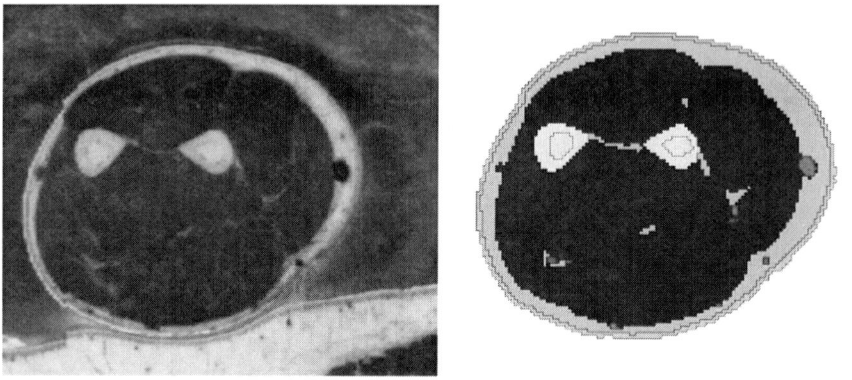

Figure 3. Example of a cropped VHD photographs before the geometric model preparation (left) and corresponding forearm slice after segmentation (right).

The Pennes' measurements (Pennes, 1948) were performed exactly "8 cm distal to tip of ulna olecranon" – or 8 cm from the tip of the elbow, with no respect to the length of the forearm. While the lengths of the forearms are not reported, the diameters of the measured forearms differed greatly (7 – 10 cm). Therefore, the measurements were likely performed at different anatomical positions. In case that the 8 cm were strictly respected, the variation in the position of the measuring probe, regarding the forearm anatomy, could be significant – up to 25% of the distance to the elbow.

The diameter of the described geometric model of the forearm at slice 50, which is in the middle of the model, is 12.4 cm. To obtain a closer match to the measured forearms, which were significantly smaller, we scale down the domain size by linearly scaling down the domain elements. In the rest of the section, we use forearm diameter of 8.3 cm, except when explicitly specifying otherwise.

4.2. Numerical Setup

We simulate the steady-state temperatures of a resting forearm near the thermo-neutral conditions. It is known that the thermo-neutral conditions of an unclothed arm with skin temperature (TS)(TS) of approximately 33 °C are obtained at ambient air temperature of 25.2 ± 1.1 °C. The initial conditions of the simulation comprise the temperatures of the modelled tissues except arteries, and are set to 36 °C. The arteries acts as heat generators, and ambient air acts as heat sink. They are set, in accordance to previously measured values, to constant values of 36.8 °C and 26 °C, respectively. During the simulation, the tissue temperatures converge with time towards their stationary values. After the simulating three hours of a resting forearm, the maximal change of temperatures during the last hour is less than 0.01 °C, which confirms that the near-steady state is reached.

Moving air is not simulated because of its significant contribution to the calculation complexity. We simulate the heat flow between the skin surface of the forearm and the ambient air using Robin boundary conditions, based on the continuity of the heat flow perpendicular to the surface of the simulated body part:

$$\lambda \frac{\partial T}{\partial \mathbf{n}} = H(T_{am} - T), \tag{30}$$

where TamTam stands for the ambient temperature. The convection coefficient H = 16 W m²/°C is determined by numerical experiments, based on simulation of the thermo-neutral conditions (Tam=26∘CandTS=33∘CTam=26∘CandTS=33∘C).

The influence of the parts of the forearm that were not simulated is considered by setting the heat flow to zero at both "ends" of the modelled arm (slice 1 and 100), indicating a thermal equilibrium between the modelled and excluded parts of the forearm.

The blood flow is modelled with the parameters set to:

$$\begin{aligned} V_1 &= 5.145 \cdot 10^{-5} \\ V_2 &= 0.322 \\ V_3 &= 0.705 \end{aligned} \quad . \tag{31}$$

The simulation setup is presented in Figure 4.

4.3. Results Of Numerical Integration

The simulated steady-state temperature field of the model slice 50 with diameter of 8 cm, after three hours of simulation, is shown in Figure 5a. The impact of the two main arteries, with constant blood temperature, is visible as peaks in the temperature profile at the locations of the arteries. The peaks are marked by two vertical lines. We can also notice that the skin temperature depends on the position because its distance from the arteries varies. The steady-state temperature profile taken from the same slice along the path of the thermocouple measurements are shown in Figure 5a by a black curve andFigure 5b by a thick black curve. The dotted curve shows the analytical solution of the Bio-heat equation (Yamazaki and Sone, 2000) for a cylindrical homogeneous model with blood flow $V = 4.7\ 10^{-4}1\ s^{-1}$, $hm = 0.62$ W/kg, skin temperature $TS = 32.5$ °C, and with the other constants set the

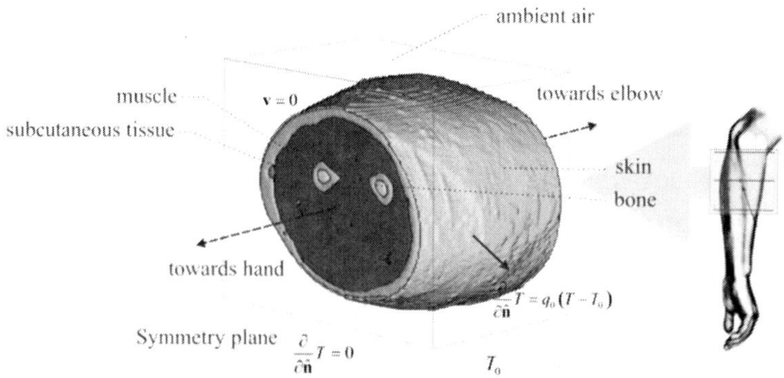

Figure 4. Computational domain

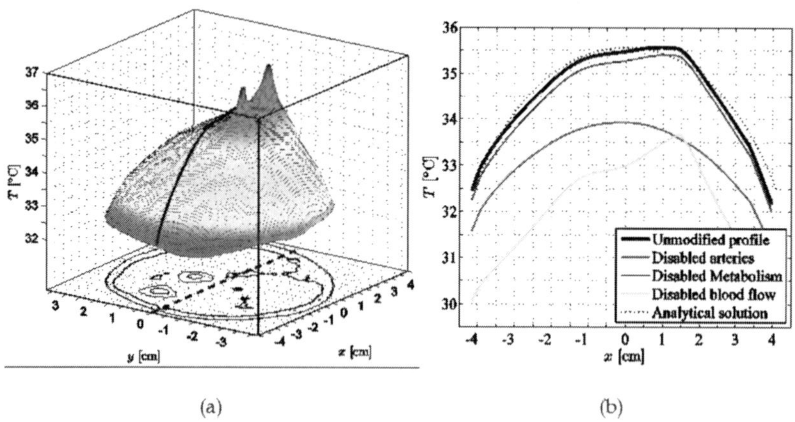

Figure 5. a) Model slice 50 and its steady-state temperature field. The simulated temperature profile along the approximate measurement path (dashed line) is emphasized with a white line. Simulated temperature profiles along the same path as in (a). The unmodified profile (thick curve) is contrasted against the analytical solution of the Bio-heat equation on a homogenous cylindrical model and the profiles acquired with modified simulation: disabled metabolism, arteries not as a heat source, and disabled blood perfusion.

same as in the simulation. We see that the closed form solution of the homogenous model can reproduce the general trends in the temperature gradients; however, it cannot reproduce the measured results because it is symmetric and uniform throughout the model.

By disabling, one at a time, the metabolism, main arteries, and blood perfusion in the simulation, we obtain the last three temperature profiles shown in Figure 5b. The blue curve shows the simulated temperature without metabolism. We can see that the metabolism minimally contributes to the final temperature. The red curve was obtained by not keeping the arteries at a constant temperature in the simulation, thus cancelling their heat generating role. The impact of the arteries not heating up the surrounding tissue makes this temperature profile stand out among the simulated profiles. It shows lower deep muscle temperature than normal, which is more symmetric. The green curve, obtained by the simulation with the arteries at a constant temperature and disabled blood perfusion, confirms that the contribution of the blood perfusion is essential. The average temperature of this profile is significantly lower with pronounced peaks corresponding to the two main arteries.

4.4. Sensitivity Analysis

The simulated steady-state solution is evaluated regarding its sensitivity. The input parameters are varied within selected ranges and the variations in the resulting temperature profiles are analysed. InFigure 6 the temperatures of the forearm after three hours of simulated resting are shown.

The temperatures are taken from the model slice 50 along the measurement path. All results are calculated based on the same geometric model with varying forearm diameter, thermal diffusivity, blood flow, and metabolism, all for +20% (dotted line) and -20% (dashed line) of their nominal values (solid line). The nominal values are taken from Table 1. The nominal value for the forearm diameter was determined to be 10 cm.

The most significant impact on the temperature profiles comes from varying the forearm dimension (Figure 6a). Larger arms result in a plateau slightly above 36.3 °C, while smaller arms are cooled more intensively, reducing the maximal temperature to 35.6 °C and with larger temperature gradients in the superficial regions of the forearm. Changes in the thermal diffusivity (Figure 6b) have a much smaller impact on the temperature profiles than the forearm dimension. The effect is also non uniform across the profile, but is rather equalizing the superficial and the deep muscle temperatures. The impact of the blood flow (Figure 6c) is again uniform across the profile, with higher forearm temperatures when the blood flow is larger. The impact of the metabolism (Figure 6d) is analogous to the one of the blood flow – greater metabolism results in higher forearm internal temperatures – but significantly smaller in scale.

4.5. Reproducibility Of Measurements

We found out that the most important parameter that influences the simulation results is the dimension of the simulated forearm. The impact of the diameter is evident from Figure 6a. Other thermal parameters, like blood perfusion and structural details, have much smaller

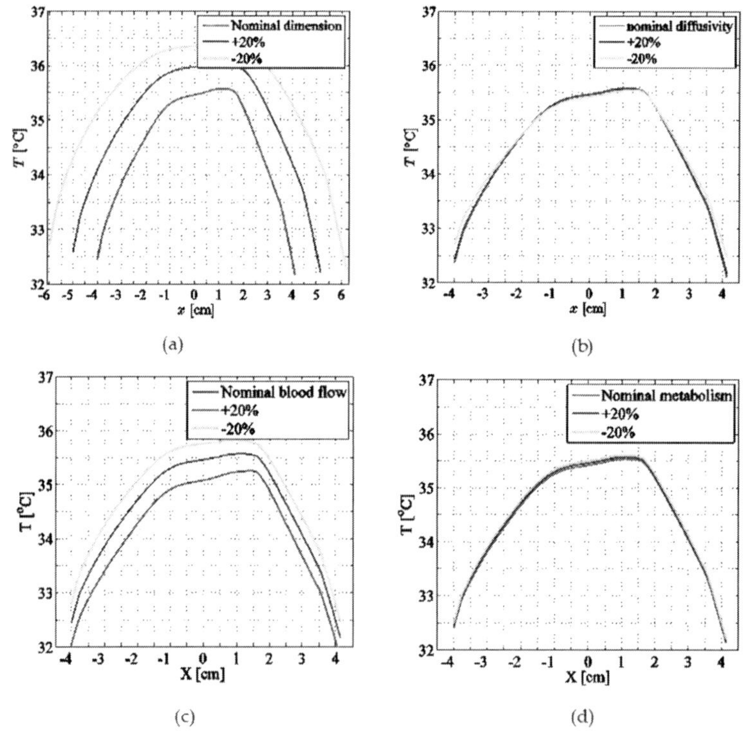

Figure 6. Temperatures from the model slice 50 along the measurement path after three hours of simulated resting forearm at a room temperature of 26 °C for various: (a) forearm diameters, (b) thermal diffusivity, (c) blood flow, and (d) metabolism. All panels feature red, blue, and green lines for the results acquired using the nominal, 20% above nominal, and 20% below nominal values of the given parameter, respectively.

impact on the variations in the simulated temperatures and have been therefore considered in the same way in all individual cases. Only two main vessels have been considered, which are anatomically quite uniform at most subjects. The remaining smaller vessels are encompassed through the blood perfusion as an average temperature regulator. The temperature gradients are very pronounced and therefore the location of the temperature probe provides the second most important contribution in the analysis of the temperature profiles. Because the anatomical probe positions are not known exactly, we try to find them to see if they can explain the unexpected variations in the measured temperature profiles.

We can deduce from Pennes work further sources of measuring errors. The positions of the measuring points and the reading of the temperature values have been with limited accuracy, which could result in errors of 0.1 °C. The dimensions of the thermocouples were up to 0.2 mm and hence, in the areas with steepest temperature gradient, near the forearm surface, an additional error could be introduced.

To reproduce the measurements, we first simulate the resting forearms with the same diameter and under the same conditions as in the real measurements. Using a matching 3-D simulation and a temperature profile measurement, we perform a search for the probe trajectory on which the simulated temperature profile fits the measured temperature profile the most. We do so by defining a fitness function as the root mean square (RMS) error between the simulated temperature profile and the measured referential profile, weighted with Gaussian function to diminish the effect of greater inaccuracies in the measurements near the forearm surface. Then, we use the fitness function in Differential Evolution (Price and Storn, 2008) – a stochastic optimization procedure – to find a probe trajectory that fits the real measurement well. The similarity between the measured and the simulated profiles for all analysed measurements is given by the Pearson correlation coefficient in Table 2. The simulated paths of the measuring probes, which provide the highest correlation between the simulated and the measured profiles, are also determined in Table 2 by two offsets and two angles to the measurement path. They can be obtained by sequentially applying translations in y and x directions and rotations in x-y and x-z planes to the trajectory of the thermocouple probe.

Two typical simulated temperature profiles for measurements 4 and 9, together with points from the original measurements, are shown in Figure 7. The central peak in the profile of measurement 4 is expected and can also be predicted by the theory. The temperature profile of measurement 9 with dual maxima can be explained by the probe moving close to and about the same distance from both arteries. The left and right shifted maxima visible

Table 2. Comparison between the measured forearm temperature profiles and the simulated temperature profiles that fit them best, for all analysed measurements.

	Subject							
	1	2	3	4	5	7	8	9
Correlation	0.988	0.995	0.993	0.980	0.997	0.989	0.980	0.991
y offset [mm]	-2.6	2.7	4.5	1.8	5.7	-0.4	2.4	1.2
z offset [mm]	-21.4	-32.7	-30.3	-9.3	-28.8	-9.0	-36.3	-4.4
Angle in *x*-*y* plane [°]	0	-0.4	-1.6	-0.4	0	0.4	0	3.5
Angle in *x*-*z* plane [°]	5.5	5.9	0	0.8	7.8	4.3	3.9	4.3

in some other measurements can be explained by the path of the temperature probe being closer to one of the arteries. The difference in the forearm dimensions results in significantly different temperatures in the central part of the forearm. As visible from Figure 7, larger forearms induce higher temperatures. We are able to reproduce the original measurements using the same geometric model for all subjects (just scaled to the measured diameter), and by introducing minor changes in the position of the measuring paths.

4.6. Summary

Instead of first simulating and then evaluating the simulated results, our study is designed in reverse direction. First, the well-known in vivo measurements of the steady-state temperatures of the proximal forearm are taken from the famous Pennes' publication; then, the experimental environment is reproduced; next, the bio heat transfer is modelled, and lastly, the measurements are simulated. The primary goal was to explain, by computer simulation, the variations in the Pennes' in vivo measurements. We have known that the measurements vary significantly between subjects and between forearm cross-sections. Several attempts have been made in the past to average the measured steady-state temperatures in order to smooth fluctuations and obtain a standardized response.

The anatomical positioning of the measuring probes is assumed to vary because of the different forearm sizes of the subjects; therefore, their possible paths have been reconstructed by searching for the simulated paths that result in the best agreement between the simulated and the measured temperature fields. Our simulated results suggest that the smoothing approach is inappropriate, and that the fluctuations are natural result and the consequence of a complex interplay between the position of the measuring probe, the anatomical position of the main arteries, the dimensions of the forearm, the blood flow, the inhomogeneity of tissues, and the environmental temperature.

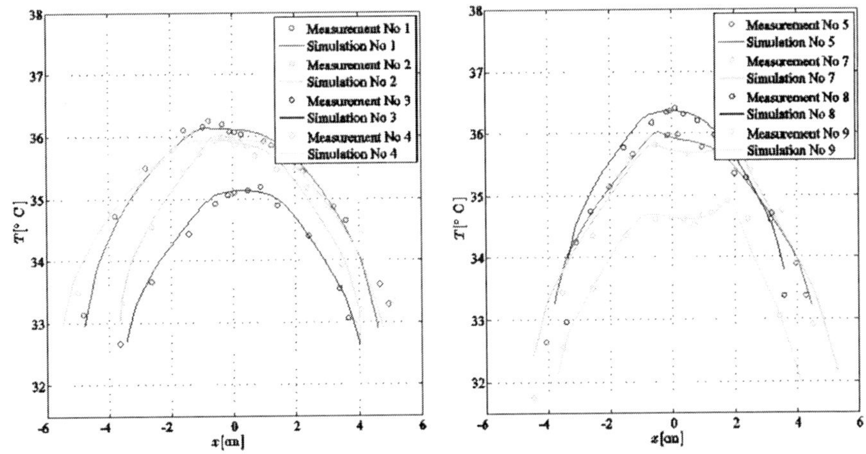

Figure 7. The best fit of the simulated temperature profiles and the corresponding measurements.

Several innovative approaches have been introduced in the simulation method. A procedure for spatial model generation based on digitized slice data has been developed. A mathematical model and a 3-D computer simulation program have been implemented for the simulation of steady-state temperatures. The heat transfer in the model of non-homogenous tissue was modelled with a well-known Bio-heat equation. The heat production by tissue metabolism was modelled using the Q10 rule. The heat exchange between the blood and the tissue was modelled as a function of the local temperature and the regional blood flow that was modelled as an exponential function fitted to experimental data. The regional blood flow was assumed to depend on the regional temperature in the same way as the mean blood flow rate depends on the skin temperature. Such an approach produced reasonable results. Nevertheless, it needs to be supported by further physiological research. The developed simulation, however, is sufficiently robust to incorporate arbitrary model functions for blood flow and metabolism. The stability of the method was confirmed by varying the simulation parameters, the initial and boundary values, and the model shape, with subsequent analysis of the results.

We have confirmed that blood flow has a significant and complex impact on the steady-state temperatures of the forearm. Lower blood flow results in lower temperature of the central part of the forearm. When the effect of blood flow on the temperature profile is increased by disabling the direct effect of the main arteries, the resulting temperature profiles match very well to the analytically produced profiles from the literature, but not well with the experimental measurements. We have shown that only when the effect of both sources, the heat distributed via blood flow and the heat emitted directly from the arteries, are considered, the measured temperature profiles can be fully explained. Furthermore, we have revealed the dependence of the temperature profile shape on the dimensions of the forearm. Smaller forearms are more likely to feature off-centre peaks in the vicinity of the main arteries than larger forearms. In larger forearms, a central plateau may be formed, masking all other potential peaks. The thermodynamic constants of the tissue and the tissue metabolism were confirmed to have a relatively minor impact on the temperature field.

Our simulation methodology has several limitations. The anatomy of the forearm differs between individuals and changes in time. Consequently, the simulation on one generalized model can differ from the experimental measurements. Minor errors in tissue segmentation and inaccurate thermodynamic constants could produce errors in the simulated results. The possible influence of different mechanisms of thermoregulation on blood perfusion has not been simulated because we assumed that their effects are stable when the forearm temperature is in its steady-state. We confirmed that the above essential findings are in close agreement with in vivo measurements, particularly if the inaccuracy of the position of the measuring path is accepted.

5. SIMULATION OF A HUMAN KNEE EXPOSED TO TOPICAL COOLING AFTER ACL SURGERY

Topical cooling after surgery or after knee injuries is often performed using gel-packs filled with gel that can be cooled well below 0 °C, yet still remaining soft and flexible; or alternatively, with cryo-cuffs, which provide a constant temperature maintained by circulating cooled liquid (Hubbard and Denegar, 2004). Desired cooling conditions are obtained by differently shaped gel-packs or cryo-cuffs, often influenced by environmental conditions and the temperatures of internal tissues (Barber, 2000). However, different tissues around the knee joint have different physical and thermodynamic properties, and respond diversely to the cooling. The knee temperature is influenced by the muscle metabolism and the circulating blood, both of which depend on the tissue temperature. Therefore, the real cooling efficacy varies in space and time in different parts of the knee. The effects of cooling on the tissue metabolism and the response of inflamed or traumatized tissue have been researched and have shown beneficial effects (Ho, *et al.*, 1994, McGuire and Hendricks, 2006). There is no generally accepted doctrine about topical cooling. Therefore, further investigation of cooling effects and different cooling regimes are needed, but these are difficult to perform by experiments alone.

In this section, we are interested in simulating possible modes of knee cooling after injury or surgery (Grana, 1994). We focused on the comparison of two different methods of topical cooling. First is the cooling with gel-pack filled with refrigerated gel (McMaster, *et al.*, 1978). The gel-pack is exposed to ambient temperature and therefore becomes less and less effective with time. Second is the cooling with a cryo-cuff filled with a liquid. The temperature of the liquid is maintained constant by an external cooling device (Shelbourne and Nitz, 1990).

Lowering the tissue temperature reduces the need for medicaments and shortens the rehabilitation period (Knight, 1985). It is not clear, however, whether different tissue regions in the knee should be cooled uniformly. Some recent findings put the benefit of topical cooling under question (Hubbard and Denegar, 2004, Knight, 1985, McGuire and Hendricks, 2006). The aim of our work is to simulate topical cooling of the knee after injury or surgery, and to calculate and show the development of temperature distribution in all tissues of the knee region.

5.1. Geometric Model Of The Knee

The knee area is cropped from the VHD male cross-section photographs number 2200 to 2390 as rectangles of 550 (width) × 610 (height) pixels. Cropped VHD slice 2301 is shown on the left in Figure 8.

The 3-D model is shown in Figure 9. Skin, joint liquid and subcutaneous tissues are not shown. The artero-lateral quadrant is removed for better visibility into the central knee region.

Some surrounding space for the protective bandage, the gel-pack, the blanket, and the ambient air is added around the 3-D knee model. The simulation environment is imitated by an isolated cube composed of x×y×zx×y×zresulting in roughly 107107 voxels independently characterized by thermodynamic properties and initial temperatures. Each voxel has a volume of 1 mm^3. The knee is covered by a 2 mm thick protective bandage and embraced by a 12 mm thick cooling layer (gel-pack or cryo-cuff). Additionally, the cooling layer is covered with a 5 mm thick isolating blanket to reduce the convection and slow down the warming from outside. The protective bandage, the cooling layer, and the isolating blanket are inserted into the model automatically by a computer program.

5.2. Numerical Setup

The boundary layers of the simulated box are held at constant initial temperatures to mimic the ambient air. The influence of the rest of the leg, which is not simulated, is managed by setting the temperature flux at the boundaries. The simulation can be carried out in 2-D with the heat flux in the axial direction set to zero. In this way, an infinitely long "knee", with homogeneous structure in the axial dimension, can be simulated. Simulation in 3-D is performed on the described model with all boundaries set to values that are similar to those of the measuring conditions. The simulation setup is presented in Figure 4.

The blood perfusion in bones, cartilage, and ligaments from the knee joint are assumed to be very small. Thus, for these tissues, ten times smaller numerical values for V0V0 are taken than for the other tissues. The blood flow is modelled using the following parameters:

$$\begin{aligned} V_1 &= 5.722 \cdot 10^{-3} \\ V_2 &= 0.187 \qquad . \\ V_3 &= 0.000 \end{aligned} \tag{32}$$

As in the case of the blood flow, the metabolism of bones, cartilage and ligaments is assumed to be very small.

The thermo-physical properties of newly introduced materials are stated in Table 3 and initial states of different regions are stated in Table 4.

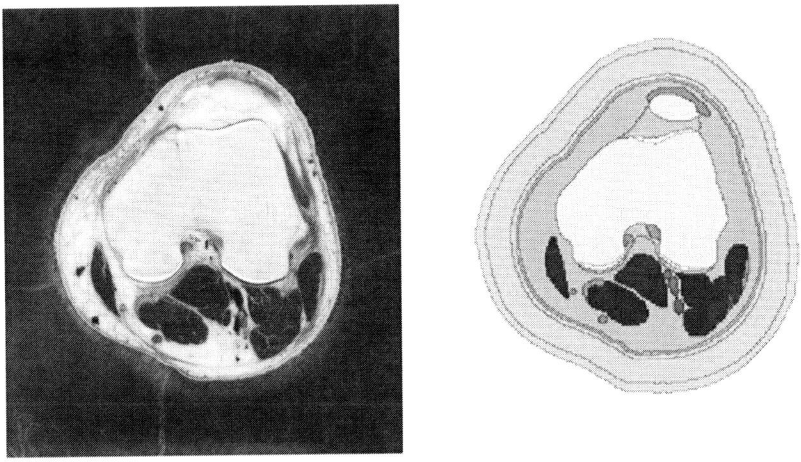

Figure 8. Example of a cropped VHD photographs before the geometric model preparation (left) and corresponding knee slice after segmentation (right).

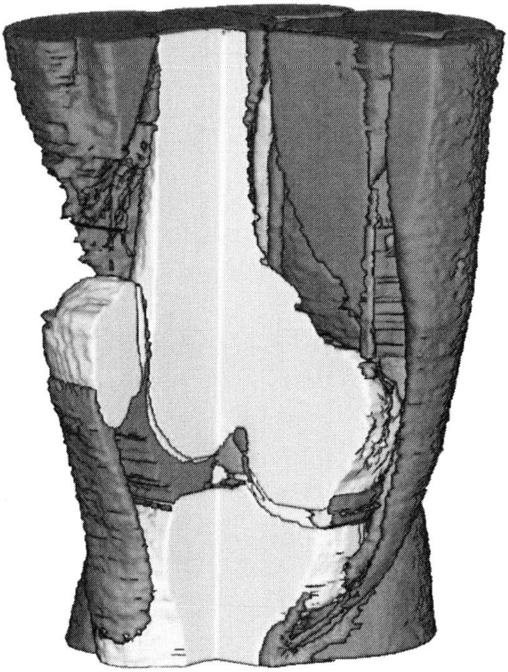

Figure 9. D geometric knee model obtained by stacking 191 consecutive VHD slices from top to bottom. The frontal quadrant, skin and subcutaneous tissue are removed for better visibility.

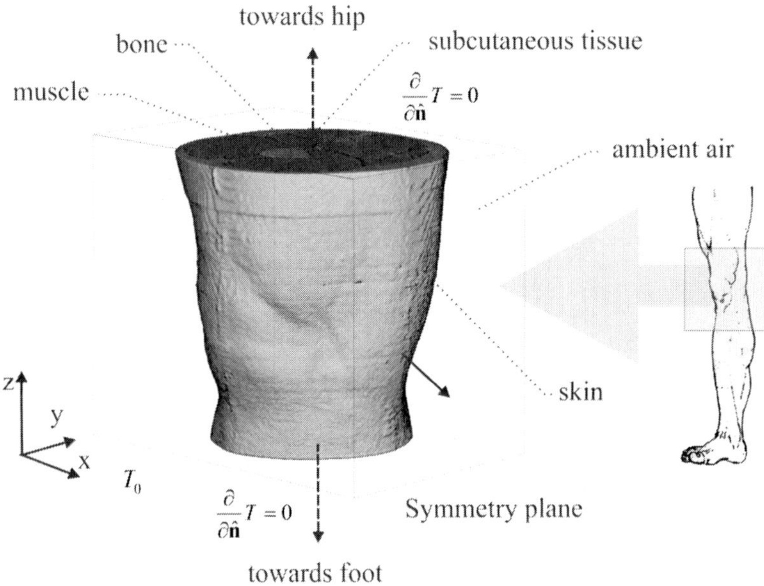

Figure 10. Computational domain.

Table 3. Thermo-physical properties of additional materials.

	$\lambda\left[\frac{\mathrm{W}}{\mathrm{mK}}\right]$	$c_p\left[\frac{\mathrm{Jkg}}{\mathrm{K}}\right]$	$\rho\left[\frac{\mathrm{kg}}{\mathrm{m}^3}\right]$
blanket	0.04	1200	150
cryo-cuff water	0.58	4204	1000
gel-pack	0.1	4000	990
bandage	0.04	1200	150
joint liquid	0.58	4204	1000
venous blood	0.67	3890	1057
arterial blood	0.67	3890	1057
blanket	0.04	1200	150

Table 4. Initial temperatures in °C.

ambient air	25	muscle	36
blanket	25	bone	36
cryo-cuff water	15*	ligament	36
gel-pack	0	cartilage	36
bandage	30	joint liquid	22
skin	35	nerve	36
subcutaneous tissue	35.6	venous blood	36
arterial blood	36.8*		

* Constant temperatures are denoted by asterisks.

5.3. Results Of Numerical Integration

First, we simulate the steady-state temperatures of a resting knee in thermo-neutral conditions (Ducharme, *et al.*, 1991) at ambient air temperature of 27 °C°C, which is equivalent to the ambient water, e.g., a water bath at 33 °C°C. A steady-state is reached after three hours of simulation, with the maximal change in the temperature near the end of the third hour being less than 0.01 °C°C.

At that state, the simulated temperatures can be recorded at arbitrary positions. For example, the temperature field over the model slice z=102z=102is shown with a 2-D surface in Figure 11. For better visibility of the temperature field, the point (0,0)(0,0) is on the right and the frontal part of the knee with the patella on the left. The temperature along the transverse axis YYfor x=106x=106is used in a later analysis. The obtained steady-state is used as an initial condition in all our subsequent simulations. The tissues nearer the patella, toward the surface, are seen to be colder than the internal part. The impact of the main knee artery with its constant blood temperature is visible as a peak in the temperature field. The location of the bones is also evident as shallow depressions in the temperature field, mainly as a result of the lower blood flow in the bones.

In Figure 12 the solid curve shows the simulated steady-state temperature obtained with the 3-D simulation. The dotted curve is obtained under the same conditions with a 2-D simulation and zero flux in the third dimension (infinite knee). The difference between the simulations is in the range of 0.5 °C°C. It can be explained by underestimating the amount of muscle tissue in the whole knee region. Slice 102 is taken, namely, from the central part of the knee with less muscle, which could contribute additional heat through the blood flow. However, the shapes of the profiles are similar.

Therefore, for the initial analysis we use 2-D simulation to reduce the simulation time.

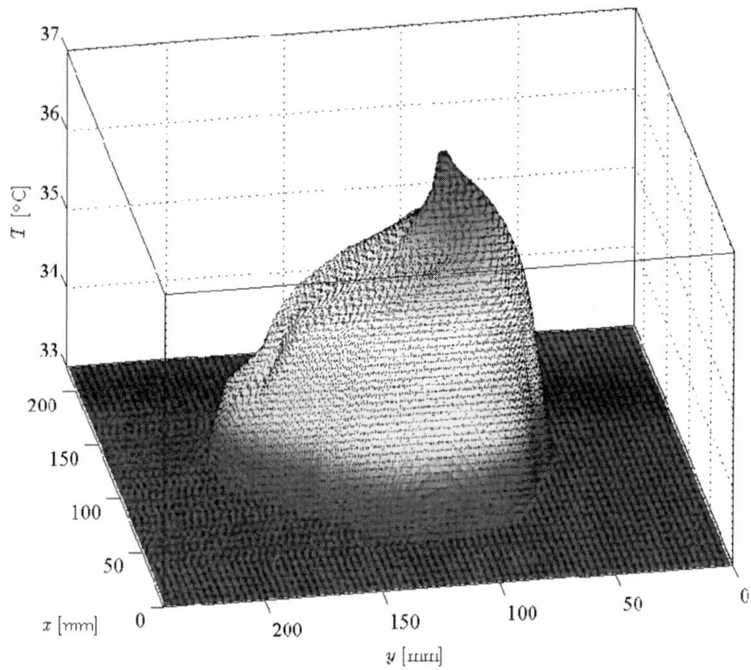

Figure 11. Steady-state temperature field over model slicez=102z=102. Point (0,0)(0,0) is in the right corner for better visibility.

We can observe and analyse arbitrary simulated points or regions of the domain. For example, in Figure 13, the steady-state temperature profiles of the knee are shown for temperatures on the transverse knee axis y y and model slice z=102z=102forx=96x=96, x=106x=106andx=116x=116. It is clear that there can be significant differences in the temperature, even for analysed points as close as 10 mm. The effect of altering the observed position in other directions is similar to that seen in Figure 13.

The stability of the simulated steady-state solution is evaluated by varying the simulation parameters. Input parameters are varied within selected ranges and the variations in the solution were analysed. InFigure 14 the simulated steady-state temperatures from model slice 102 on the transversal axis yyfor x=106x=106are shown for various knee dimensions, diffusion constants, blood flow and metabolism, all differing for +20 % and -20 % of their nominal values.

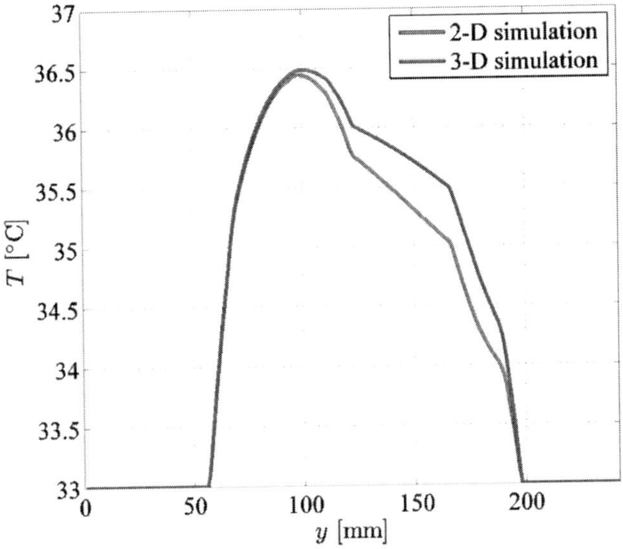

Figure 12. Simulated steady-state temperature profiles by thermo-neutral conditions at the slice z=102z=102along the transverse knee axis yyforx=106x=106, for 2-D and 3-D simulation.

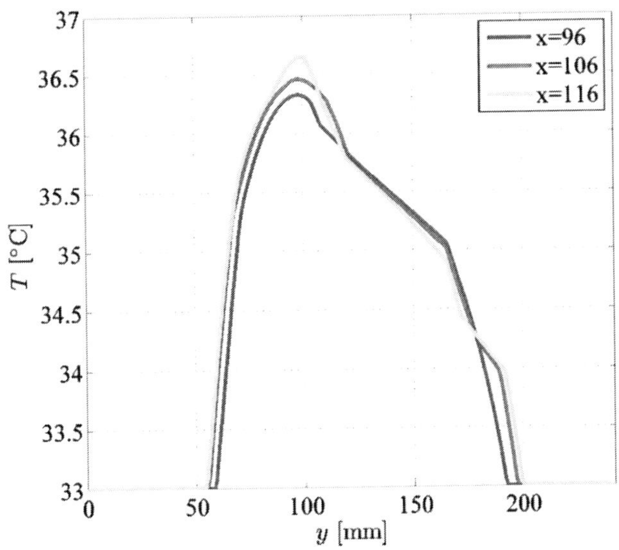

Figure 13. Simulated steady-state temperature profiles by thermo-neutral conditions on the transverse knee axis YYand model slice z=102z=102forx=96x=96, x=106x=106andx=116x=116.

The most important impact on the temperature profiles seen in Figure 14 arises from varying the knee dimension (Figure 14a). Larger knees result in a temperature plateau slightly above 36.7 ℃°C, while smaller knees are cooled more intensively to the central temperature of about 36 ℃°C, with larger temperature gradients in the superficial regions of the knee.

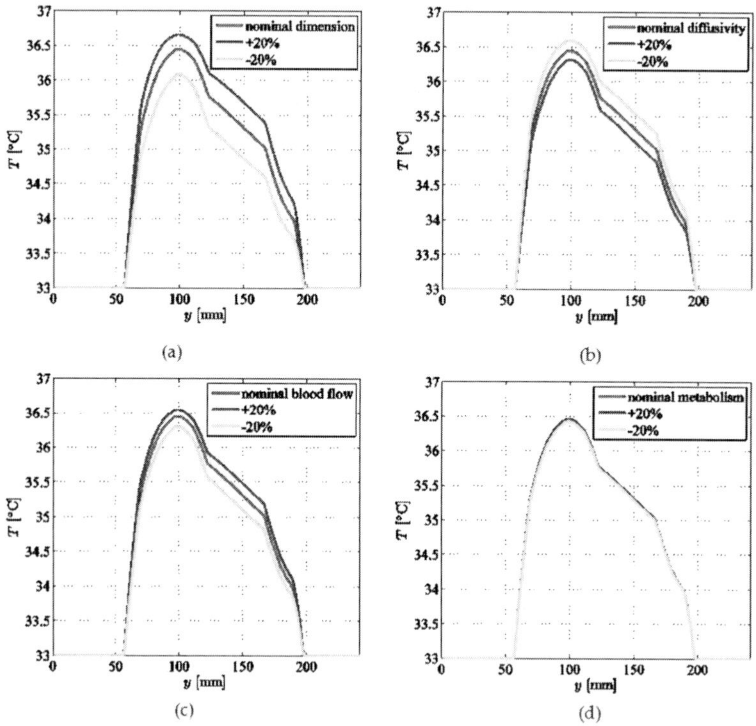

Figure 14. Simulated steady-state temperatures from model slice z=102z=102and the transversal axis YYfor x=106x=106for various: (a) knee dimension, (b) thermal diffusivity, (c) blood flow and (d) metabolism. In all Figure -panels nominal values are red, 20 % above nominal values are blue, and 20 % below nominal values are green.

Changes in thermal diffusivity (Figure 14b) have a smaller impact on the temperature profiles than changes in the knee dimensions, and in the opposite direction, i.e., larger diffusivity constants result in lower central temperatures.

The impact of blood flow (Figure 14c) is similar to that of knee dimensions, with higher internal temperatures arising from greater blood flow. However, an important difference is that the shapes of the temperature profiles remain unchanged with variations of the blood flow.

The impact of metabolism (Figure 14d) is analogous to that of the blood flow – increased metabolism results in higher internal temperatures, but its impact is so small that it can be neglected in our experiments.

Simulated steady-state temperatures of a resting knee under thermo-neutral conditions (TN) are used as an initial condition for all further simulations. First, we simulated a naked knee in steady-state at ambient air temperature of 25 ℃°C. Then, we simulate a two-hour period of an arthroscopic operation, during which the knee joint is washed out by sterilized water at 22 ℃°C, and therefore cooled. During the following two-hour period we simulate the temperature evolution in the operated knee while it is resting and is covered by a blanket, and therefore warming. Finally, we simulate a subsequent two-hour postoperative topical cooling.

5.4. Washing Out During Arthroscopy

During arthroscopic reconstruction of ligaments, the central part of the knee is washed by sterilized water at 22 ℃°C. The water circulates in the space around the femoral intercondylar notch, normally filled by the joint liquid. The initial temperatures are taken from the steady-state of a naked knee at an ambient temperature of 25 ℃ °C. The temperature of the joint liquid is fixed at 22 ℃°C. The temperature profile after two hours of washing out are shown in Figure 15 for the same plane as before; that is, along the YYaxis forx=106x=106, but for the 7 mm higher slice atz=95z=95, because this is nearer the actual position of our measuring probes for validation of the simulated results. The internal knee temperature decreases significantly, maximally by more than 14 ℃°C in places that have direct contact with the washing water, as can be seen from the temperature profiles in Figure 15.

5.5. Resting After Arthroscopy

Next, we simulate the two hour period immediately after surgery. The knee is resting and is covered with a blanket in a room temperature of 25 ℃°C. The evolution of the temperatures in voxel(106,128,95)(106,128,95), which is at the level of the femoral intercondylar notch in the central part of the knee, and in voxel(52,126,95)(52,126,95), which is nearer to the knee surface, 1 cm below the skin in the subcutaneous tissue, are shown in Figure 16. Note that arbitrary voxels can be selected for the analysis. The knee is initially colder in the central region because of the previous washing out with cold water. During resting, its temperature increases and approaches the steady-state, with colder regions nearer the skin.

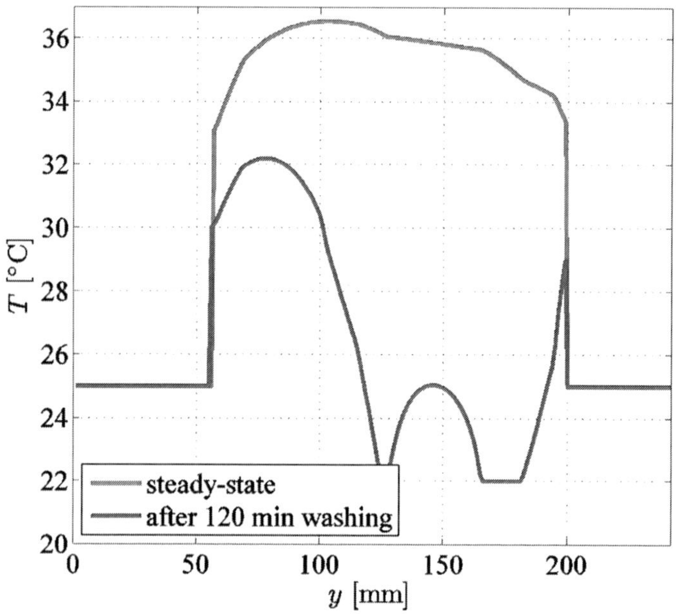

Figure 15. Steady-state temperature profiles for a naked knee and after a two hour washing out during surgery, for the model slice z=95z=95and the transversal axis yyforx=106x=106.

5.6. Postoperative Topical Cooling

Finally, we simulate postoperative topical cooling with two different cooling methods, gel-pack and cryo-cuff. In both cases, the knee is bound with a protective blanket surrounded by ambient air at 25 ℃°C.

In Figure 17 the temperatures of voxels (106,128,95) (106,128,95) and (52,126,95)(52,126,95) are shown for the two-hour simulated period, for cooling with a gel-pack (initial temperature 0 ℃°C) and a cryo-cuff (water with constant temperature of 15 ℃°C). The effects of the two methods on the knee temperatures are quite different.

When cooling with gel-pack, the temperature of the voxel (106,128,95)(106,128,95) in the central knee region initially slightly increases because of the arterial blood perfusion and metabolism, and the weak influence of the initial cooling. After 5 minutes, the temperature of the voxel starts to decrease. However, after 40 minutes, the gel-pack has received enough heat from the knee surface and the ambient air, allowing the inner knee temperature to increase during the second part of the cooling period. For the voxel (52,126,95)(52,126,95) in the subcutaneous tissue, 10 mm below the skin, the temperature first sharply decreases; the effectiveness of the gel-pack then becomes weaker, the voxel starts warming up, and after 120 minutes reaches almost 36.0 $℃$°C.

The cooling with a cryo-cuff is found to be more effective. It induces lower tissue temperatures, even if the temperature of the cooling liquid is as high as 15 $℃$°C. In the initial phase, both voxels experience the same cooling rate as the one with the gel-pack. However, there is no subsequent increase in temperature because the cryo-cuff is a constant heat sink which gradually cools the knee. After two hours of cooling, the near surface voxel reaches a temperature of 27 $℃$°C. In the same way, but with smaller intensity, the inner voxel is cooled to 33 $℃$°C.

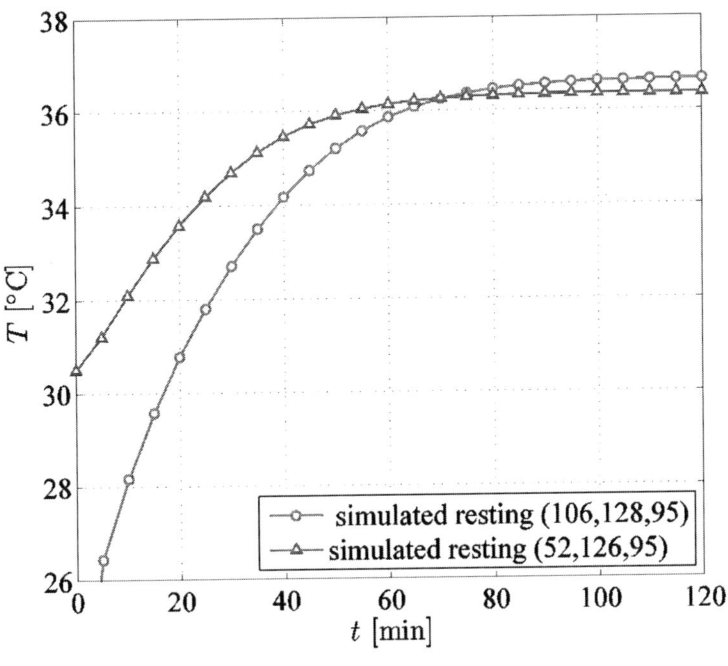

Figure 16. Simulated temperature evolution in two resting hours after surgery for voxels (106,128,95)(106,128,95) (the central part of the knee) and (52,126,95)(52,126,95) (in the subcutaneous tissue).

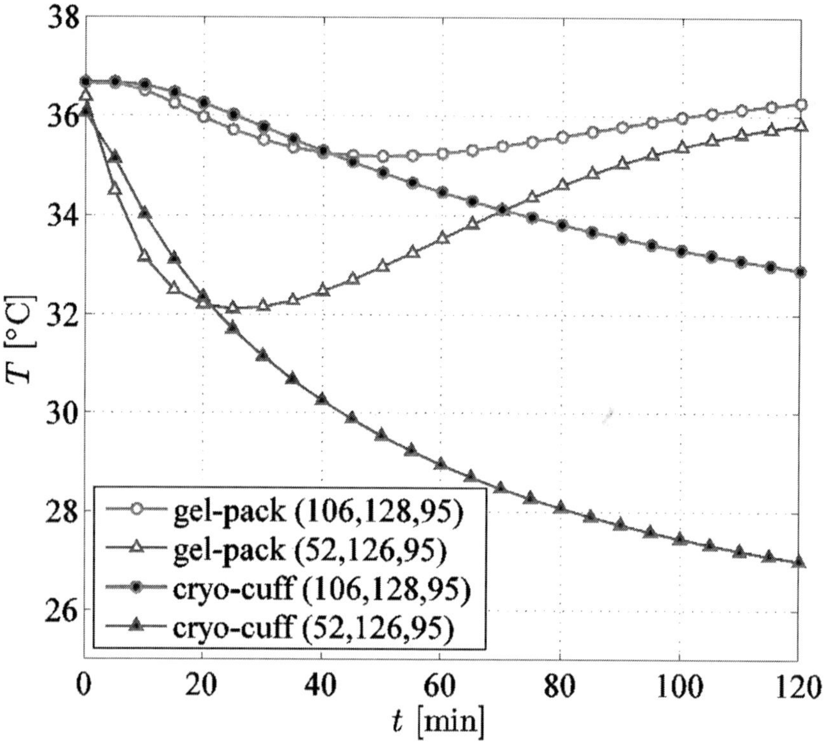

Figure 17. Simulated temperature evolution in voxels (106,128,95)(106,128,95) and(52,126,95)(52,126,95), as a function of time for cooling with gel-pack and cryo-cuff.

In the case of topical cooling with gel-pack, the heat of fusion should also be simulated, which is necessary for the transition between aggregate states. Obviously, for a crushed gel-pack, a significant part of the heat is needed for such a transition, which would prolong the effective cooling time. This phenomenon has not been incorporated in our mathematical model. Instead, we increase the heat capacity c c of the gel-pack to recognize and account for this behaviour. The simulated results show that the topical cooling with a cryo-cuff provides more constant lowering of the temperatures in the whole region of the knee. Cooling with gel-packs is less stable; consequently, they should be changed every half an hour in order to be effective.

In Figure 18 temperature profiles after one hour of simulated cooling with gel-pack and cryo-cuff are shown for a cross-section from the patella to the lateral side of the knee on our standard axis, i.e., along the YY(anteroposterior) axis for x=106x=106on the model slicez=95z=95.

After one hour of simulated cooling, the gradients in the temperature profiles were much more pronounced than in the initial state. When cooling with gel-pack, the temperature of the outer knee layers at the skin level remained cooled to 32 ℃°C, and in the centre of the knee to 36 ℃°C. The peaks in the tissue temperatures around y=100y=100result from the simulated heat conduction from the middle popliteal artery. Significantly lower temperatures are observed for the cooling with cryo-cuff, even though its constant temperature was as high as 15 ℃°C.

Given the above results, it would be interesting to test the effectiveness of a simple cooling with ambient air. The knee would remain uncovered and exposed to the ambient air temperature of approximately 20 ℃°C. We expect, from the results under thermo-neutral conditions, that the skin temperature would be about 25 ℃°C, which could lower the temperatures inside the knee. This could be tested preliminarily using the proposed simulation method.

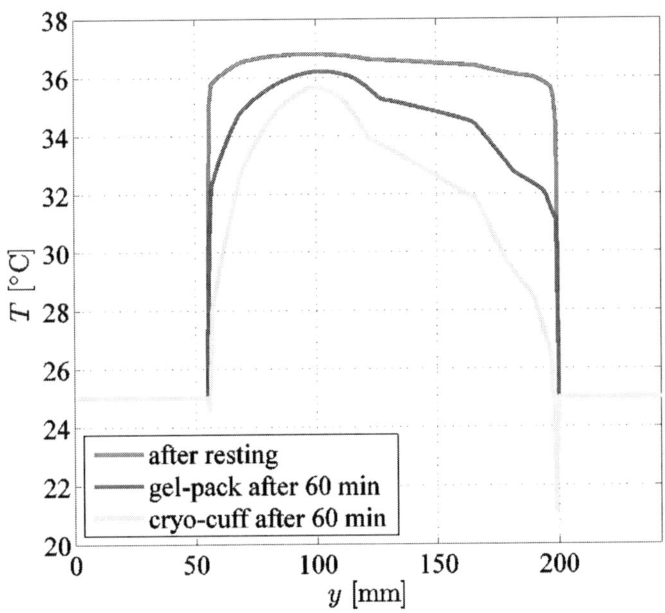

Figure 18. Simulated initial knee temperature profile after resting and temperatures after one-hour period of cooling with the gel-pack and cryo-cuff along the YYaxis for x=106x=106andz=95z=95.

5.7. Validation Of Results

To evaluate the simulated results, we made two control measurements of the knee temperature following surgery in a room with constant ambient temperature of 25 ℃°C. We measured the temperature of the knee covered with a blanket during the two-hour resting period immediately after surgery; then, during the next two-hour period in which topical cooling with a gel-pack was applied.

Two small thermistors were placed into the knee in thin sterile tubes (Foley-catheter with temperature sensor, 3 mm, Ch8-thermistor; Curity, Degania Silicone Ltd., Degania Bet, Israel). Similar tubes, without thermistors, are ordinarily inserted for wound drainage following surgery. The thermistors were connected to a registration device for continuous measurement with a sample rate of 0.1 Hz and resolution of 0.01 ℃°C.

The first thermistor was placed in the centre of the knee near the voxel (106,128,95) (106,128,95), and the second approximately 1 cm below the skin in the subcutaneous tissue, near the voxel (52,126,95) (52,126,95). The measurements were approved by the Slovenian State Medical Ethics Committee and the patient gave written informed consent prior to participation.

In Figure 19 the simulated and measured temperatures are shown for the two-hour resting period after washing out during arthroscopic surgery. The knee was wrapped in a protective blanket at ambient temperature of 25 ℃°C under the same conditions as in the simulation.

The simulated temperature evolution for the resting period shows very good agreement with the measured values for the test point (106,128,95)(106,128,95) in the central part of the knee. However, the simulated rate of cooling in point (52,126,95)(52,126,95) in the subcutaneous tissue was much smaller in the initial phase than the one obtained by the measurements. One of the possible reasons is the fact that we did not simulate the cold washing out inlet that also cooled the surrounding tissue from the skin to the central part of the knee. The subcutaneous thermistor was placed in such cooled environment which could then exhibit faster warming than in our simulation. In Figure 20 the simulated and measured temperatures are shown for two hours of cooling with a gel-pack, which follows immediately after the resting period. The initial temperature of the gel-pack was 0 ℃°C and the ambient temperature 25 ℃°C. The knee was bound with elastic bandages approximately 2 mm thick, surrounded with fixed gel-packs and wrapped in a protective blanket.

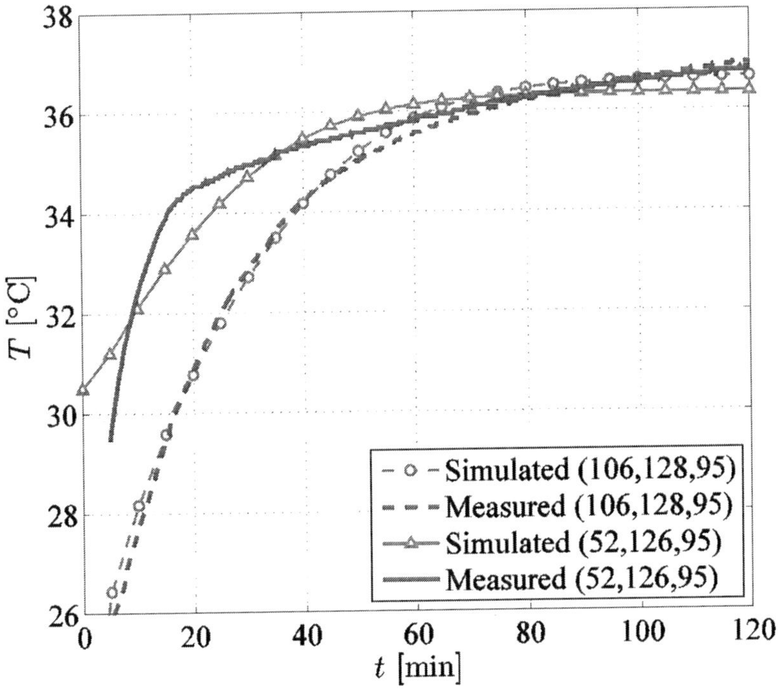

Figure 19. Measured and simulated temperature evolution of voxels (106,128,95)(106,128,95) and (52,126,95)(52,126,95) in a two hour resting period after arthroscopic surgery.

The simulated evolution of temperature for the cooling period shows good agreement with the measured values for both test points. The simulated rates of warming in the second hour are slightly greater than by measurement. One of the possible reasons is the incomplete mathematical model that does not include the heat of fusion for the gel-pack. In fact, all the thermodynamic characteristics of the gel were not available, and we just took some approximate values provided by the supplier. Another possible reason is inaccurate measurement. We did not collect detailed data during the measurements, e.g., the wetness of the protective bandage, which could have a significant impact on the cooling intensity

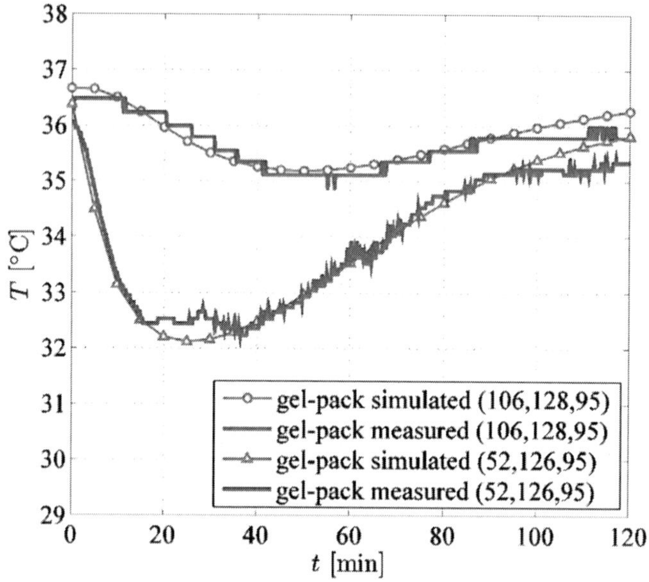

Figure 20. Measured and simulated temperature evolution of voxels (106,128,95)(106,128,95) and (52,126,95)(52,126,95) for cooling with a gel-pack.

5.8. Summary

We demonstrated a practical application of the simulation program for two different methods of topical postoperative cooling of a knee. We simulated topical cooling of a knee with a gel-pack. The inner knee tissues reached their lowest temperature in 40 minutes. For continued effective cooling, the gel-pack has to be replaced. The topical cooling with a cryo-cuff was more effective. We simulated situations with relatively small cooling rates to analyse the small influences of the blood flow and metabolism. The thickness of a protective bandage and isolating blanket, or their thermal conductivity, together with the cooling temperature, can be used to regulate the cooling intensity. The results have been validated by experimental measurements of knee temperatures. The simulation results confirm that the model and methodology used are appropriate for the thermal simulation of bio-tissues.

We have shown that the blood flow has a significant and complex impact on the stationary state temperatures, and the gradient of temperature change in the subcutaneous tissue and in the tissues nearer the central part of the knee. Lower blood flow results in linear temperature profiles with larger gradients, for example in bones. The dimensions of the knee are also important factors influencing the temperature distributions and gradients. Temperatures in smaller knees will differ more from the arterial blood temperature than those in larger knees. At the same time, the temperature gradient will be much greater in smaller knees. Thermal constants and metabolism have a relatively minor impact on the temperature field.

The simulated results were visualized and compared with measured values. Good agreement was obtained, leading to the conclusion that the model and method used in our simulation are appropriate for such medical simulations. Although there are not many studies of knee temperatures measured in vivo, and the measuring conditions are often not described in sufficient detail, we have also run our simulation software with the initial conditions described in some published measurements. In (Martin, *et al.*, 2001) the temperature in the lateral gutter of the knee decreased by 4 *℃*°C, one hour after knee arthroscopy. Similar values have been obtained by our simulation program.

The method described has several limitations. The 3-D knee model used in our simulation is not complete, as only a small part of the leg above and below the knee was included. The remaining part of the leg that was not included was compensated by a constant flux in boundary conditions. Spatial models differ with different persons and with time, and consequently, the simulated results can differ. Minor errors in tissue segmentation and inaccurate thermodynamic constants could also produce small errors in the simulated results. Moving air was compensated by an artificial thin layer of air with non-constant temperature. Incorporation of a fluid-flow model would be needed for even more accurate results. The possible influence of blood perfusion by different regulatory mechanisms has not been simulated. Personal regulatory mechanisms have not been included in the simulation model, but could easily be incorporated. Listed limitations could have some impact on the simulated temperatures, but the essential findings agree remarkably well with existing experimental in vivo measurements.

6. SIMULATION OF HUMAN HEART COOLING DURING SURGERY

During longer surgery the human body and the heart have to be managed appropriately in order to slow down their vital functions. One of the options for lowering metabolic requirements is the cooling of the body and the heart, e.g., by whole body hypothermia or infusion of a cold solution in coronary arteries (cardioplegia). For even better cardiac cooling, a method of topical cooling of the heart is sometimes applied (Olin and Huljebrant, 1993) by using packs filled with cooled gel or ice slush.

The efficiency of cooling is reflected in the temperature distribution within the heart, which is hard to determine. In vivo temperature measurements are invasive and, if applicable, limited to a few test points. These issues can be circumvented to some extent by physical modelling and computer simulation. Appropriate simulation tools provide insight in the temperature distribution through the whole heart volume and its neighbourhood, and therefore, various preliminary evaluations and analyses of different cooling regimes can be assessed.

Initial results on the simulation of heart cooling based only on diffusion have been reported in (Trunk, *et al.*, 2003). In this section, we present simulation results obtained through much more complex physical model, incorporating, beside diffusion, also advection, bio-heat sources and fluid flow. Such a model stands for more realistic variant of already developed simulation tools (Šterk and Trobec, 2005, Trobec, *et al.*, 1998).

Our final goal is to implement an accurate and reliable simulation framework that enables simulations of various diagnostics or therapeutic medical procedures. Simulation of different protocols of myocardial protection is taken as a test-case because they incorporate most of the physical phenomena during surgery that are related to the bio-heat transport. In this final section we focus on topical cooling as an adjunct to cardioplegia in classical non-beating heart operation (Allen, et al., 1992). During cardiac surgery, local hypothermia is reported to be successful in slowing down the tissue metabolism (Birdi, et al., 1999). This is achieved by infusing a cold cardioplegic solution into the coronary vessel system. Alternatively, topical cooling is applied with a cold saline solution, sometimes mixed with ice slush, which is distributed around the heart to further lower the temperature of the heart tissue. In non-perfused tissues, ischemia could provide deleterious effects on the cells, finally producing necrosis. Lowering of tissue temperature lengthens the time in which the ischaemic damage to the cells is reversible, which can prolong the safe time available for cardiac surgery. To provide adequate protection to the whole heart tissues, the heart has to be cooled down as uniformly as possible (Gonzales, *et al.*, 1985). The anterior part of the heart, especially the right ventricular wall is likely to remain outside the topical cooling solution during the operation, because it represents the uppermost part of the heart in supine patient. The importance of uniform temperature distribution in heart tissues comes also from the fact that areas with different temperatures could change myocardial contraction velocity (Riishede and Nielsen-Kudsk, 1990) and trigger cardiac arrhythmias. Some previous computer simulations have already shown irregular temperature distribution in the human heart during topical cooling (Geršak, *et al.*, 1997).

Opinions about the cryo-therapeutic effects on the heart during cardiac surgery and during post-surgery rehabilitation period are still controversial. Furthermore, recent findings put the real benefit of topical cooling under question (Cordell, 1995). The optimal temperature at which the heart tissues should be cooled down and the time intervals of optimal cooling are still not clearly defined, probably because of the lack of in vivo measurements and difficult detailed post-surgery patients screening.

The aim of our work is to accurately simulate topical cooling during induced cardiac arrest in a typical heart operation. We want to use 3-D geometric model with all details, including coronary vessels. The desired result is the temperature distribution in the heart tissues as evolving in the time of surgery. The topical cooling is mimicked by submerging the heart in a cooling liquid. We are particularly interested in the influence of the liquid dynamics on the cooling performance.

Detailed quantitative validation of the results is not the main goal of our investigation. We are more concerned about the adequacy analysis of the presented physical model for simulation of heat transfer phenomenon during open heart surgery. For the sake of simplicity and transparency of results, we assume some simplifications, which can be declared as limitations of our study.

First, we simulate only the topical cooling and do not take into account the heart cardioplegia and the whole body hypothermia. It allows us to be concentrated only on a single cooling modality. The impacts of the remaining cooling methods remain to be studied in the future. The whole body hypothermia could be simulated by setting the domain boundary conditions to lower temperatures. The cardioplegia could be approximated by setting the temperature of blood in the coronary vessels to a constant value with Dirichlet boundary conditions on the vessel walls, assuming well mixed blood with a significant flow. Next, we simplify the domain boundary conditions to the Dirichlet, although the cooling would affect the body as well; however, the body temperature is assumed to remain constant through physiological regulatory loops. Next, because we assume well mixed air we neglect the natural convection of air in the operating room. The ambient temperature is set to constant value. Finally, blood in heart chambers is assumed to be steady.

6.1. Geometric Model Of A Heart

VHD male cross-section photographs, numbered 1350 to 1505, are selected for the construction of the geometric model. *Z* axis of the VHD is parallel to the heart axis, pointing approximately from the apex towards the atria and the base of the heart. The selected cross-sections are cropped to the size of 512 (width)× 512 (height) pixels. Different tissues are classified (segmented) manually with an assistance of expert anatomist. The space occupied by voxels not belonging to the heart tissues or major vessels (aorta, vena cava, pulmonary vessels) are replaced either by voxels of air, water or the box wall. The final 3-D model used in the simulation is shown in Figure 22.

During the tissue segmentation, a few problems that could deteriorate the desired spatial accuracy of the heart model have to be resolved. The details of the heart structures have to be preserved as much as possible. The accuracy of the simulations relies on the resolution of the geometric model, which serves as a substrate for the simulation. For example, it is impossible to distinguish between the epicardium and pericardium on the VHD photographs; therefore, these two tissues are treated as a single tissue. The wall of the large vessels originating in the heart chambers is also not distinguishable from the neighbouring pericardium. The myocardial tissue contains several layers of muscular fibres that run in different directions. This could influence the heat propagation through the tissue, if different directions are examined; however, these layers could not be distinguished on the original cross-sections. Thus, the myocardium is classified as a homogenous mass of muscular tissue. The left ventricle, usually much larger in a heart in function, is quite small in the VHD dataset, evidently contracted and being in a systole like phase. On the contrary, the right ventricle is huge and elliptically shaped with an extremely thin wall. The endocardium, which normally lines the entire interior of the heart chambers, is microscopically thin, so it could not be differentiated in our model.

The border between two tissues was sometimes difficult to be determined from the VHD photographs because of the similar pixel colours. In such cases, the borders were drawn subjectively according to the anatomical knowledge. The majority of problems are detected in the area of the pulmonary and other large vessels at the base of the heart. These vessels can be identified only with concurrent examination of neighbouring slices and subsequent merging of multiple-slice information. Also, the pericardium is discontinued in some places, particularly on the posterior surface of the heart, because it is not distinguishable from the surrounding tissues on the posterior side of the heart. In such cases, the borders are drawn subjectively with some variations, taking into account a few consecutive neighbouring cross-sections. The small coronary vessels are also hard to be identified. The left and right coronary arteries, the circumflex artery and the largest veins are visible and one can follow such vessels from their origin to the periphery. On the periphery, coronary arteries and veins sometimes lie in close proximity and look like a single large vessel. Some vessels do not have a uniform course in the 3-D view. There are discontinuations that occurred because the vessels are not recognized correctly on every cross-section. Some further artefacts are also detected, like spots painted on some cross-sections as vessels, but actually not corresponding to vessel tissues. This results in a single pixel or a group of pixels apart from the coronary vessels. Therefore, the coronary vessel system was re-segmented again in a separate phase.

The segmentation errors of the generated spatial heart model have to be corrected, either manually or automatically by a computer program. A spatial editor program is implemented to allow the user to easily spot and correct the smaller failures in the generated 3-D models. The spatial editor is also used for several automatic modifications of the 3D-model, e.g., a uniform reduction of the model size by a user supplied factor *n*. Another function is designed for finding small openings or discontinuities in a selected tissue (exposing other tissues) and repairing them by adding small portions of missing tissue onto adequate places of the cross-sections. This function was mostly used to fill in small holes in the pericardium.

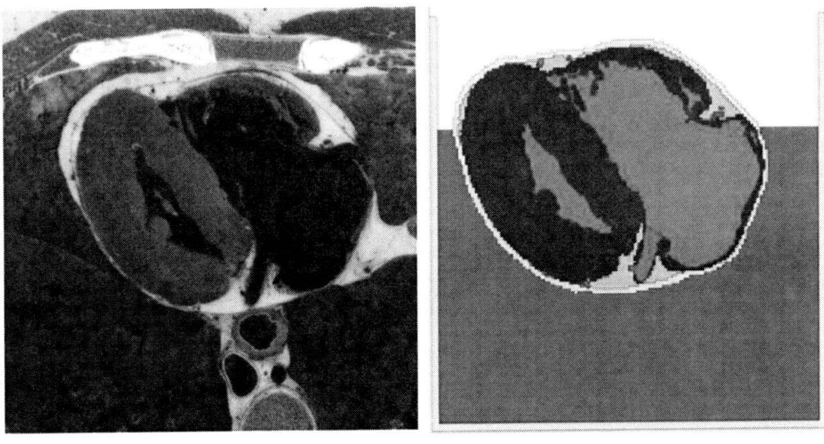

Figure 21. Example of a cropped VHD photographs before the geometric model preparation (left) and corresponding heart slice after segmentation (right).

6.2. Numerical Setup

The thermal conditions during cardiac surgery are imitated with an isolated cube large enough to contain the simulated heart model. The computational domain is shown Figure 22. The lower part of the cube represents a cooling container filled with a cooling liquid that is initially set to a temperature of 0.2 °C, assumed to be cooled via ice slush. The container walls are kept at a constant temperature of 36.5 °C, simulating the body temperature. The initial temperature of the heart is also set to

36.5 °C. The upper part of the cube represents the open thorax surrounded by air at room temperature. The surrounding air is assumed to be well mixed and, as such, is kept on a constant temperature of 20 °C. The cube is composed of 163×169×163 voxels of 1 mm³ representing a 3-D box with size x = 163 mm, y = 196 mm, z = 163 mm independently characterized by specific thermodynamic constants, defined earlier in Table 1 (Section 2) and Table 3 (Section 5).

The blood flow in a Bio-heat equation (7) is modelled using the following parameters:

$$V_1 = 5.145 \cdot 10^{-5}$$
$$V_2 = 0.322$$
$$V_3 = 0.705. \tag{33}$$

The cooling liquid is assumed to be water with thermal expansion coefficient β=4·10−4β=4·10−4 and viscosityμ=1.2·10−3Pasμ=1.2·10−3Pas. The heat transfer in water is simulated without bio heat sources (10).

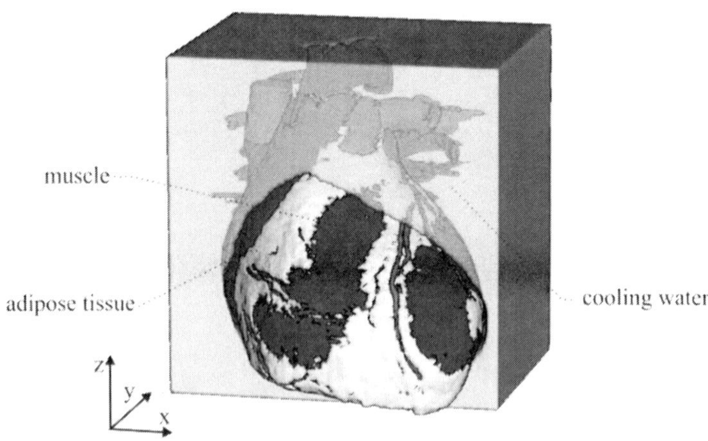

Figure 22. Computational domain of simulated heart.

6.3. Results

The present analysis is focused on the detailed assessment of the impact of fluid flow on the cooling process of the heart. Note that in cases described in previous sections the cooling is modelled only as a diffusive process.

The heat flux from the boundaries and the heart warms the surrounding water. The temperature dependent density governed by the Boussinesq approximation induces non-uniformly distributed volumetric body force in the Navier Stokes momentum equation, thus, a velocity field develops. The phenomenon is also known as natural convection (Amimul, 2011). The moving fluid extensively affects the cooling process through the advection (second term in the right side of equation (7)) term. In this section, we show the importance of heat advection. Simulation time of 10 minutes is considered, as the characteristics relevant to our discussions already develop in that time. The results are presented in terms of velocity profiles and temperature contour plots on *x-y* cross-section at $z = 80$ mm in four characteristic time instants t = [60, 240, 420, 600] s, and in terms of cross-lines in *x-y* plane for four different cross-sections z = [30, 60, 90, 120] mm at the end of simulation (600 s).

We compare two different scenarios; first, with fluid flow and second, without fluid flow. The main difference is in the treatment of the cooling liquid. The first case represents complex simulation with diffusion, advection and bio-heat sources, while the second case is similar to previously presented simulations of arm and knee. Such a cooling method with the "still" water can be approximated by cold gel packs or ice slash packs, where the flow is not possible.

In Figure 23 the temporal development of the setup with fluid flow is presented. The effects of advection, due to the natural convection, are clearly visible. Near the boundaries, the water is warmer as it is constantly heated through the domain walls and the heart. The heat flux generates upward fluid flow that eventually turns inwards as it reaches the boundary of the free liquid media. The consequent fluid flow significantly increases the heat transport from the domain boundaries, as well as from the heart into the cooling water. Note that in the bottom of the domain, local flow instabilities can be identified. The local bursts of upward flow result from instable buoyant force distribution. The warmer fluid pressures upward and only a small perturbation is needed to start a local upward burst of fluid, resulting in a local vortex. The induced circulation significantly increases the heat transfer between the walls and the heart.

In Figure 24 the simulation without fluid flow is presented. There is no advection and the heat extracted from the heart is transported only by means of diffusion, which is much slower in comparison with the advection. There is significantly less heat transported from the walls and, therefore, more heat is extracted from the heart before water reaches steady state temperature. The steady state, although not simulated in the present work, is also reached considerably later if the advection is not in place, as the heat flow is much slower (see water temperature on the last Figure panel - 600 s). In the case with diffusion only, the water plays an additional role of an insulator between the heart and the boundary walls.

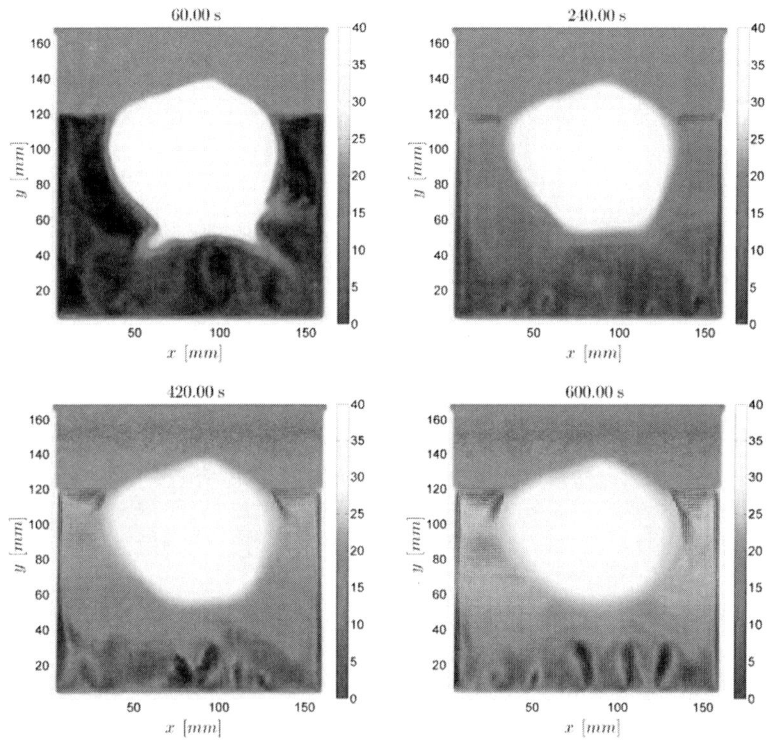

Figure 23. Temperature and velocity fields with natural convection shown on the cross-section at z = 80 mm at four time instants [60, 240, 420, 600] s.

Finally, the whole domain is examined by considering temperature cross-section plots after the simulated 600 s. Three cross lines in x-y plane at z = [30, 60, 90, 120] mm for both simulated cases (with and without natural convection) are shown in Figure 25. These additional representations confirm the earlier conclusions. The water retains its initial temperature for much longer period of time in the case without flow. The immediate effect, visible after ten minutes of simulation, is lower heart temperature in the case without advection (curves marked by circles).

6.4. Summary

We confirmed that the fluid flow plays a crucial role in the process of topical cooling of the heart. The natural convection in the cooling liquid increases the dynamics of the heat transport. The advection increases the flux from the domain boundaries, as well as from the

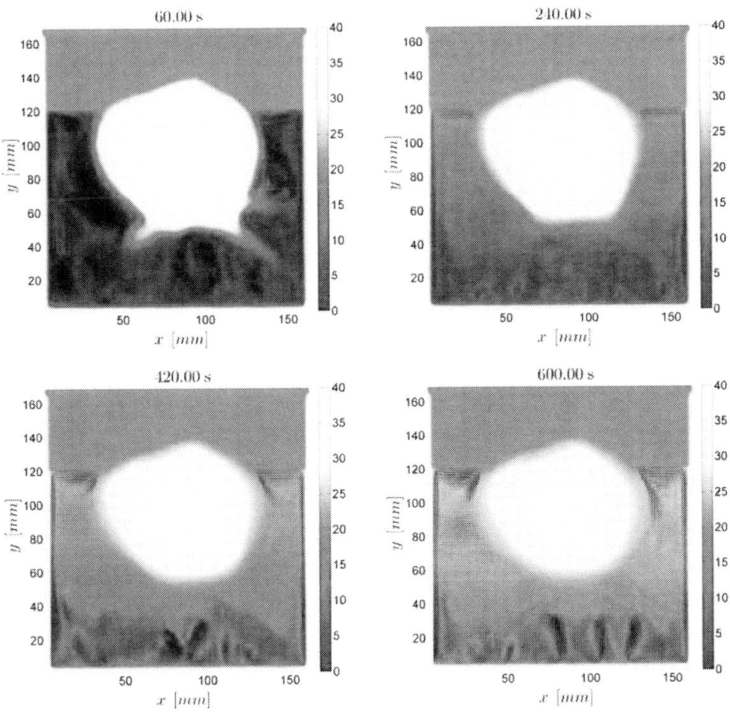

Figure 24. Temperature fields without natural convection shown on the cross-section at z = 80 mm at four time instants [60, 240, 420, 600] s.

heart walls. In the presented cases, the walls are kept at a constant temperature. Hence, the increased flux primarily affects the heating of the water. Consequently, the cooling process is less effective. If the domain boundaries would be kept at lower temperatures or kept isolated, e.g., by heart transplantation, the advection induced cooling would be much more pronounced. Alternatively, the water could be externally cooled or completely replaced after certain period of time to keep the cooling process more effective; however. The above cooling options are beyond the scope of our work and remain to be further investigated. In the case of omitted advection, the impact of cold "still water" lasts longer, resulting in lower simulated temperatures of the heart. We expect that new standards for the cooling protocols have to be developed in the future, supported also with the results and methodology presented in our work.

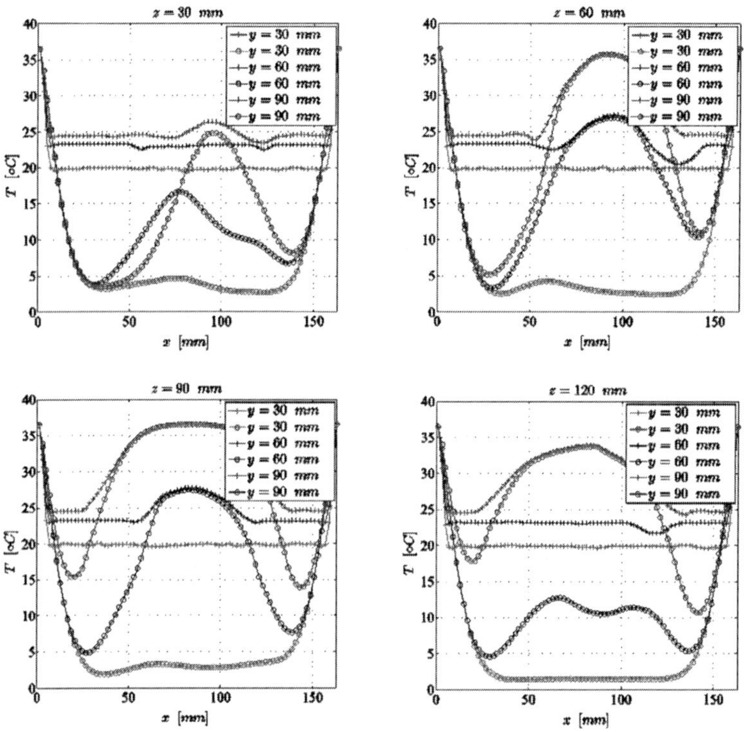

Figure 25. Comparison of temperature distribution at different cross-sections after 10 minutes of simulation: complete simulations (including advection) are marked with plus signs, simulations without advection are marked by circles.

7. DISCUSSION AND CONCLUSIONS

We developed and implemented a physical model for the solution of the Bio-heat equation described in Section 2. The model incorporates heat diffusion, heat advection, heat transfer through blood perfusion and heat generation by tissue metabolism or other bio sources. A corresponding mathematical model and a computer simulation program have been implemented. Some important extensions have been introduced, in particular, an inhomogeneous spatial model composed of tissues with different characteristics, and modelling of the blood perfusion and heat sources as functions of the surrounding tissue temperature.

The heat produced by tissue metabolism was modelled using the Q10 rule, while the heat exchange between the blood and the tissue was modelled as a function of the regional blood flow and the local temperature. The regional blood flow was modelled as an exponential function fitted to experimental data. It was assumed to depend on the regional temperature in the same way as the mean blood flow rate depends on the skin temperature. Such an approach produced reasonable results. Nevertheless, it needs to be supported by further physiological research. The developed simulation, however, is sufficiently robust to incorporate arbitrary model functions for blood flow and metabolism. The stability of the method was confirmed by varying the simulation parameters, the initial and boundary values, and the model shape, with subsequent analysis of the results.

The construction of geometrical model, described in Section 3, was based on VHD slices. The models were composed of equal voxels, which results in a regular spatial discretization. Consequently, an explicit finite difference scheme has been developed and optimized for this purpose. A general methodology for the development of 3-D models, based on digitalized slice data, is described and applied for generating our spatial geometric models.

We have shown that the computational complexity of the simulation is proportional to the geometric resolution. The long run times of the presented 3-D simulation, which was in the range of several hours per run, constitute a technical limitation. The initial simulations were done in 2-D on a single model slice, which gives first merits about the temperature of deep tissue. Although the 2-D simulation times were shorter by several orders of magnitude, there were only minor differences in the steady-state temperature profiles. The final results presented in this chapter have been obtained by the 3-D simulation. Simpler models with lower geometric resolution could not reproduce vessels, which significantly deteriorate the accuracy of the simulation results. Regarding the available measuring equipment and environmental clinical conditions (e.g., variations in room temperature), the simulation accuracy should be approximately 0.1 °C.

We also implemented a parallel version of the proposed method, which runs efficiently on 16 or more connected computers in a computing cluster. In this way, the computation time can be shortened significantly. This approach enables the solution of several millions of equations for each time-step. Besides the simulated results, the computer simulations also support the development of new ideas and theories because a lot of "unexpected" situations can be explained. The proposed method can be applied in investigations of a variety of living tissues.

The solution of the Bio-heat equation was used in three different cases in the field of medicine. In the first case, described in Section 4, our primary goal was to explain, by computer simulation, the variations in the Pennes' in vivo measurements of the steady-state temperatures along the transverse axis of the proximal forearm. Assuming that the anatomical positioning of the measuring probes varied because of the different forearm sizes of the subjects, we have reconstructed their possible paths by searching for the simulated paths that result in the best agreement between the simulated and the measured temperature fields. Several attempts have been made in the past to average the measured steady-state temperatures in order to smooth fluctuations and obtain a standardized response. Our simulated results suggest that this approach is inappropriate, and that the fluctuations are natural result and the consequence of a complex interplay between the position of the measuring probe, the anatomical position of the main arteries, the dimensions of the forearm, the blood flow, the inhomogeneity of tissues, and the environmental temperature.

We have confirmed that blood flow has a significant and complex impact on the steady-state temperatures of the forearm. Lower blood flow results in lower temperature of the central part of the forearm. When the effect of blood flow on the temperature profile is increased, by disabling the direct effect of the main arteries, the resulting temperature profiles match very well to the analytically produced profiles from the literature, but not well with the experimental measurements. We have shown that only when the effect of both sources, the heat distributed via blood flow and the heat emitted directly from the arteries, are considered, the measured temperature profiles can be fully explained. Furthermore, we have revealed the dependence of the temperature profile shape on the dimensions of the forearm. Smaller forearms are more likely to feature off-centre peaks in the vicinity of the main arteries than larger forearms. In larger forearms, a central plateau may be formed, masking all other potential peaks. The thermodynamic constants of the tissue and the tissue metabolism were confirmed to have a relatively minor impact on the temperature field.

In the second test case, described in Section 5, we demonstrated a practical application of our simulation program in topical postoperative cooling of a knee with two different cryotherapeutic methods. First, we simulated the topical cooling of a knee with a gel-pack. The inner knee tissues reached their lowest temperature in 40 minutes. For continued effective cooling, the gel-pack has to be replaced. The topical cooling with a cryo-cuff was more effective. We simulated situations with relatively small cooling rates in order to be able to analyse the small influences of the blood flow and metabolism. We observed that various thicknesses of a protective bandage and isolating blanket, or variations in their thermal conductivity and cooling temperature, can be used to regulate the cooling intensity.

The results have been validated by experimental measurements of knee temperatures. The simulation results confirm that the model and methodology used are appropriate for the thermal simulation of bio-tissues. We have shown that the blood flow has a significant and complex impact on the stationary state temperatures, and on the gradient of temperature change in the subcutaneous tissue and in the tissues nearer the central part of the knee. Lower blood flow results in linear temperature profiles with larger gradients, for example, in bones. The dimensions of the knee are a very important factor influencing its temperature distribution and gradient. Temperatures in smaller knees will differ more from the arterial blood temperature than those in larger knees. At the same time, the temperature gradient will be much greater in smaller knees. Thermal constants and metabolism have a relatively minor impact on the temperature field.

We confirmed that fluctuations in temperature profiles are natural results and a consequence of the complex interplay between the positions of the measuring probes, the dimensions of the investigated body part, the anatomical positions of the main arteries and bones, and the environmental temperature. The fluctuations and augmented peaks in the temperature profiles can be explained by our simulation results, since the temperature changes in the centre of the knee are influenced by the vicinity of the main leg artery with constant blood temperature. The steepest, almost linear, gradients in the area of the subcutaneous tissues are observed in the outer parts of the simulated temperature profiles, particularly in cases with lower cooling temperatures. The temperature plateaus, measured earlier by other investigators, in the inner knee region with increased dimensions of the knee, were also demonstrated by our simulation.

The simulated results were visualized and compared with measured values. Good agreement was obtained, leading to the conclusion that the model and method used in our simulation are appropriate for such medical simulations. Although there are not many studies of knee temperatures measured in vivo, and the measuring conditions are often not described in sufficient detail, we have also run our simulation software with the initial conditions described in some published measurements. In (Martin, *et al.*, 2001) the temperature in the lateral gutter of the knee decreased by 4 *°C*°C, one hour after knee arthroscopy. Similar values have been obtained by our simulation program. More detailed control studies should be done to compare experimentally measured and simulated results in order to fine-tune the simulation method.

The last test case, described in Section 6, stands for the most complex computation in the present chapter. It incorporates all physical model elements, including incompressible fluid flow. We confirmed that the fluid flow plays a crucial role in the process of topical cooling of the heart. We showed that, under simulation assumptions, the cooling process is less effective if natural convection is considered. In the case with no advection, the impact of rigid cold substance lasts longer, resulting in lower simulated temperatures of the heart. However, if the cooling liquid would be externally regulated, the results might be different. The advection increases the heat transfer and if appropriate cooling water management would be chosen, the cooling efficiency might be improved.

We are aware that our simulation methodology still has several limitations. A separate study should be devoted to the validation of simulated results; however, in many cases, in vivo measurements are not applicable or significantly limited. The anatomy of the human body differs between individuals and changes in time; consequently, the simulation on a single generalized model can differ from the experimental measurements. Some kind of personalized approach is foreseen in the future, also because of possible influences of different mechanisms of thermoregulated blood perfusion. When known, these mechanisms could be easily incorporated into the simulation model. Finally, the moving air and the blood in heart chambers were approximated without fluid flow in order to simplify the simulation. Although all these limitations could have some impact on the simulated temperatures, the essential findings are in close agreement with in vivo measurements.

Future work includes the simulation of the temperature field evolution in time and analysis of the possible impact of the measuring probes, surgical instruments and other medical equipment on the temperature fields. Using an approach analogous to the presented one, other parts of the human body could also be simulated and their temperature fields could be analysed. The proposed methodology could be useful for researchers in developing methods for accurate prediction of the temperature distribution in inhomogeneous tissues. Computer simulations could help in the study of various medical applications. It has been proved that the techniques described can be used to predict the temperature distribution in deep tissues, at any body point and time of interest, and for different environmental conditions.

For future development of the numerical approach we will consider the promising local meshless methods, like Radial Basis Function Collocation Method (LRBFCM) (Šarler, 2007) or Diffuse Approximate Method (DAM) (Prax, *et al.*, 1996). The meshless methods are not limited to the equidistant uniform nodal distribution as the FDM is. The meshless approach would improve the accuracy near irregular geometric model boundaries. The DAM shows good stability and accuracy behaviour on non-regular nodal distributions (Trobec, *et al.*, 2012), and consequently, is convenient for treatment of the presented bio-medical computations. The adaptive nodal distribution (Kosec and Šarler, 2011) would also improve the treatment of areas with high gradients, e.g., the interfaces between different materials with discontinues in thermo-physical properties. The meshless methods have been recently under intense research in fluid flow context, as well. The local pressure velocity coupling procedure in combination with LRBFCM has been proposed (Kosec and Šarler, 2008a, Kosec and Šarler, 2008b). The local fluid flow solution procedure is simpler to paralelize as it does not require anykind of global treatment.

The spatial models and the simulation programs, presented in this work, are available from the authors for research purposes that could contribute to any kind of knowledge in the area of biomedical simulation.

ACKNOWLEDGEMENT

We thank the staff of the University Medical Centre Ljubljana, Department of Cardiovascular Surgery and Department of Traumatology, Slovenia. We also acknowledge the financial support from the state budget by the Slovenian Research Agency under the grant P2-0095.

REFERENCES

1. M. J. Ackerman, 1998The visible human projectProc IEEE, 86505511
2. S. G. Akl, 1997Parallel Computation: Models and Methods,Prentice Hall, New Jersey.
3. B. S. Allen, G. D. Buckberg, E. R. Rosenkranz, W. Plested, J. Skow, E. Mazzei, R. Scanlan, (1992), Topical cardiac hypothermia in patients with coronary disease, J Cardiovasc Surg, 104 626631 .
4. A. Amimul, 2011Convection and Conduction Heat TransferIntech, Rijeka.
5. S. N. Atluri, S. Shen, (2002), The Meshless Method, Tech Science Press, Encino.
6. S. N. Atluri, 2004The Meshless Method (MLPG) for Domain and BIE Discretization, Tech Science Press, Forsyth.
7. F. A. Barber, 2000A comparison of crushed ice and continuous flow cold therapy,Am. J. Knee Surg., 1397101

8. H. Barcroft, O. G. Edholm, 1942The effect of temperature on blood flow and deep temperature in the human forearmJ. Physiol., 102520

9. H. Barcroft, O. G. Edholm, 1946Temperature and blood flow in the human forearmJ. Physiol. (Lond.), 104366376

10. P. Bernardi, M. Cavagnaro, S. Pisa, E. Piuzzi, 2003Specific Absorption Rate and Temperature Elevation in a Subject Exposed in the Far-Field of Radio-Frequency Sources Operating in the 10-900-MHz Range,IEEE Trans Biomed Eng, 50295304

11. R. Białecki, E. Divo, A. Kassab, 2006Reconstruction of time-dependent boundary heat flux by a BEM-based inverse algorithmEngineering Analysis with Boundary Elements30767773

12. I. Birdi, M. Caputo, G. Underwood, D. Angelini, A. J. Bryan, 1999Influence of normothermic systemic perfusion temperature on cold myocardial protection during coronary artery bypass surgery,Cardiovasc Surg, 7369374

13. H. F. Bowman, E. G. Cravalho, M. Woods, 1975Theory, measurement, and application of thermal properties of biomaterials, Annual Review Biophysics and Bioengineering, 44380

14. B. L. Bullock, P. P. Rosendahl, 1992Pathophysiology, Adaptations and Alteration in Function,Lippincott,

15. C. K. Charny, 1992Mathematical models of bioheat transfer, in: Y.I. Cho (Ed.), Advances in Heat Transfer, Academic Press, New York.

16. W. Chen, 2002New RBF collocation schemes and kernel RBFs with applications, Lecture Notes in Computational Science and Engineering, 267586

17. K. E. Cooper, O. G. Edholm, R. F. Mottram, 1955The blood flow in skin and muscle of the human forearmJ. Physiol. (Lond.), 128258267

18. A. R. Cordell, 1995Milestones in the developement of cardioplegia, Ann Thorac Surg, 60793806

19. R. Dann, J. Hoford, S. Kovacic, M. Reivich, R. Bajcsy, 1989Evaluation of elastic matching system for anatomic (CT, MR) and functinal (PET) cerebral images,Journal of Computer Assisted Tomography13603611

20. J. Detry-M, R. , G. L. Brengelmann, L. B. Rowell, C. Wyss, 1972Skin and muscle components of forearm blood flow in directly heated man, J. ApplPhysiol., 32506511

21. M. B. Ducharme, J. Frim, 1988A multicouple probe for temperature gradient measurements in biological materials,J. Appl. Physiol., 6523372342

22. M. B. Ducharme, W. P. Van Helder, M. W. Radomski, 1991Tissue temperature profile in the human forearm during thermal stress at thermal stability,J. Appl. Physiol., 7119731978

23. M. B. Ducharme, P. Tikuisis, 1994Role of blood as heat source or sink in human limbs during local cooling and heating,J. Appl. Physiol., 7620842094

24. M. B.. Ducharme, J. Frim, 1988A multicouple probe for temperature gradient measurements in biological materials,J. Appl. Physiol., 6523372342

25. J. H. Ferziger, M. Perić, 2002Computational Methods for Fluid DynamicsSpringer, Berlin.

26. B. Geršak, T. Gabrijelčič, R. Trobec, B. Slivnik, 1997Temperature distribution in human heart during hypothermic cardioplegic arrestCor Europeum, 6172176

27. A. C. Gonzales, T. A. Brandon, R. L. Fortune, S. F. Casano, M. Martin, D. L. Benneson, 1985Acute right ventricular failure is caused by inadequate right ventricular hypothermia, J. Thorac Cardiovasc Surg, 89386399

28. W. A. Grana, 1994Cold modalities, in: J. C. DeLee and D. Drez (Eds.), Orthopaedic Sports Medicine, Principles and Practice, WB Saunders, Philadelphia.

29. J. Harris, F. Benedict, 1918A biometric study of human basal metabolism,Proc Natl Acad Sci USA, 4370373

30. S. S. Ho, M. N. Coel, R. Kagawa, A. B. Richardson, 1994The effects of ice on blood flow and bone metabolism in knees,Am. J. Sports Med., 22537540

31. C. P. Hong, 2004Computer Modelling of Heat and Fluid Flow Materials Processing, Institute of Physics Publishing Bristol.

32. T. J. Hubbard, C. R. Denegar, 2004Does cryotherapy improve outcomes with soft tissue injury,J. Athl. Train, 39278279

33. E. J. Kansa, 1990Multiquadrics- a scattered data approximation scheme with application to computational fluid dynamics, part I, Computers and Mathematics with Applications, 19127145

34. K. L. Knight, 1985Cryotherapy: theory, technique and physiology,Chatanooga Corporation, Chattanooga.

35. G. Kosec, B. Šarler, 2008aSolution of thermo-fluid problems by collocation with local pressure correctionInternational Journal of Numerical Methods for Heat and Fluid Flow, 18868882

36. G. Kosec, B. Šarler, 2008bLocal RBF collocation method for Darcy flowCMES: Computer Modeling in Engineering & Sciences, 25197208

37. G. Kosec, B. Šarler, 2011H-Adaptive Local Radial Basis Function Collocation Meshless MethodCMC: ComputersMaterials & Continua, 26227254

38. G. R. Liu, (2003), Mesh Free Methods, CRC Press, Boca Raton.

39. A. G. Malan, R. W. Lewis, 2011An artificial compressibility CBS metod for modelling heat transfer and fluid flow inheterogeneous porous materials, International Journal for Numerical Methods in Engineering, 87412423

40. S. S. Martin, K. P. Spindler, J. W. Tarter, K. Detwiler, H. A. Petersen, 2001Cryotherapy: an effective modality for decreasing intraarticular temperature after knee arthroscopy,Am. J. Sports Med., 29288291

41. R. L. Martino, C. A. Johnson, E. B. Suh, 1994Parallel Computing In Biomedical-Research,Science, 265902908

42. E. C. Mc Elfresh, P. J. Kelly, 1974Simultaneous determination of blood flow in cortical bone, marrow, and muscle in canine hind leg by femoral artery catheterization,Calcified Tissue International, 14301307

43. D. A. Mc Guire, S. D. Hendricks, 2006Incidences of frostbite in arthroscopic knee surgery postoperative cryotherapy rehabilitation, Arthroscopy, 221141e11141e6

44. W. C. Mc Master, S. Liddle, T. R. Waugh, 1978Laboratory evaluation of various cold therapy modalities,Am. J. Sports Med., 6291294

45. W. M. Moses, G. L. Geddes, 1990Measurement of the thermal conductivity of equine cortical bone, American Society of Mechanical Engineers, 17185188

46. C. L. Olin, I. E. Huljebrant, 1993Topical Cooling of the Heart- A Valuable Adjunct to Cold Cardioplegia, Scand. J. Thorac. Card., 415558

47. M. N. Őzisik, 2000Finite Difference Methods in Heat TransferCRC Press, Boca Raton.

48. H. H. Pennes, 1948Analysis of tissue and arterial blood temperature in the resting human forearm,J. Appl. Physiol., 193122

49. E. Ponder, 1962The coefficient of thermal conductivity of blood and various tissues, JGen Phisiol, 45545551

50. H. F. Poppendiek, R. Randall, J. A. Breeden, J. E. Chambers, J. R. Murphy, 1967Thermal conductivity measurements and predictions for biological fluids and tissuesCryobiology3318327

51. C. Prax, H. Sadat, P. Salagnac, 1996Diffuse approximation method for solving natural convection in porous MediaTransport in Porous Media22215223

52. K. V. Price, R. Storn, 2008Differential evolution: A simple evolution strategy for fast optimization, Dr. Dobb's Journal, 221824

53. C. I. Proulx, M. B. Ducharme, G. P. Kenny, 2003Effect of water temperature on cooling efficiency during hyperthermia in humans,J. Appl. Physiol., 9413171323

54. M. M. Rahman, T. Siikonen, 2008An artificial compressibility method for viscous incompressible and low Mach number flowsInternation Journal of Numerical Methods in Engineering, 7513201340

55. M. Rappaz, M. Bellet, M. Deville, 2003Numerical Modelling in Materials Science and Engineering, Springer-Verlag, Berlin.

56. L. Riishede, F. Nielsen-Kudsk, 1990Myocardial effects of adrenaline, isoprenaline and dobutamine at hypothermic conditions,Pharmacology and Toxicology, 66354360

57. M. Shashkov, 1996Conservative Finite-Difference Methods on General GridsCRC, Boca Raton.

58. K. Shelbourne, P. Nitz, 1990Accelerated rehabilitation after anterior cruciate ligament reconstruction,Am. J. Sports Med., 18292299

59. D. H. Silverthorn, 2001Human Physiology, An Integrated Approach,Prentice-Hall, New Jersey.

60. G. D. Smith, 2003Numerical Solution of Partial Differential Equation- Finite Difference Method, Oxford University Press, Oxford.

61. H. Sugimoto, W. W. Monafo, 1987Regional Blood Flow in Sciatic Nerve, Biceps Femoris Muscle, and Truncal Skin in Response to Hemorrhagic Hypotension,J. Trauma, 2710251030

62. B. Šarler, 2007From global to local radial basis function collocation method for transport phenomenaSpringer, Berlin, 257282

63. M. Šterk, R. Trobec, 2005Biomedical Simulation of Heat Transfer in a Human HeartJ. Chem. Inf. Mod., 4515581563

64. P. Tikuisis, M. B. Ducharme, 1991Finite-element solution of thermal conductivity of muscle during cold water immersion,J. Appl. Physiol., 7026732681

65. S. Traivivatana, P. Boonmarlert, P. Thee, S. Phongthanapanich, P. Dechaumphai, 2007Combined adaptive meshing technique and characteristic-based split algorithm for viscous incompressible flow analysisApplied Mathematics and Mechanics2811631172

66. R. Trobec, B. Slivnik, B. Geršak, c. T. Gabrijelčič, 1998Computer simulation and spatial modelling in heart surgeryComputers in Biology and Medicine, 28393403

67. R. Trobec, M. Šterk, S. Almawed, M. Veselko, 2008Computer Simulation of Topical Knee CoolingComputers in Biology and Medicine3810761083

68. R. Trobec, M. Šterk, B. Robič, 2009Computational complexity and parallelization of the meshless local Petrov-Galerkin methodComputers & Structures878190

69. R. Trobec, G. Kosec, M. Šterk, B. Šarler, 2012Comparison of local weak and strong form meshless methods for 2-D diffusion equationEngineering Analysis with Boundary Elements36310321

70. P. Trunk, B. Geršak, R. Trobec, 2003Topical Cardiac Cooling-Computer Simulation of Myocardial Temperature Changes, Computers in Biology and Medicine, 33203214

71. C. B. Wenger, L. A. Stephenson, M. A. Durkin, 1986Effect of nerve block on response of forearm blood flow to local temperature,J. Appl. Physiol., 61227232

72. E. H. Wissler, 1998Pennes' 1948 paper revisited, J. Appl. Physiol., 853541

73. L. C. Wrobel, 2002The Boundary Element Method:Applications in Thermo-Fluids and Acoustics, John Wiley & Sons, West Sussex.

74. F. Yamazaki, R. Sone, 2000Modulation of arterial baroreflex control of heart rate by skin cooling and heating in humans,J. Appl. Physiol., 88393400

75. K. Yue, X. Zhang, F. Yu, 2007Simultaneous estimation of thermal properties of living tissue using noninvasive methodInt J Thermophysics, 2814701489

9

CHAPTER

Experimental and Modeling Investigation of the Effect of Air Preheat on the Formation of No_x in an RQL Combustor

G. S. Samuelsen[1], J. Brouwer[1], M. A. Vardakas[1,2], J. D. Holdeman[3]

[1]*UCI Combustion Laboratory, University of California, Irvine, Irvine, CA 92697-3550, USA*

[2]*Solar Turbines, San Diego, CA, USA*

[3]*National Aeronautics and Space Administration, Cleveland, OH 44135, USA*

ABSTRACT

The Rich-burn/Quick–mix/Lean-burn (RQL) combustor concept has been proposed to minimize the formation of oxides of nitrogen (NO_x) in gas turbine systems. The success of this low-NO_x combustor strategy is dependent upon the links between the formation of NO_x, inlet air preheat temperature, and the mixing of the jet air and fuel-rich streams. Chemical equilibrium and kinetics modeling calculations and experiments were performed to further understand NO_x emissions in an RQL combustor. The results indicate that as the temperature at the inlet to the mixing zone increases (due to preheating and/or operating conditions) the fuel-rich zone equivalence ratio must be increased to achieve minimum NO_x formation in the primary zone of the combustor. The chemical kinetics model illustrates that there is sufficient residence time to produce NO_x at concentrations that agree well with the NO_x measurements. Air preheat was found to have very little effect on mixing, but preheating the air did increase NO_x emissions significantly. By understanding the mechanisms governing NO_x formation and the temperature dependence of key reactions in the RQL combustor, a strategy can be devised to further reduce NO_xemissions using the RQL concept.

1 INTRODUCTION

The Rich-burn/Quick-mix/Lean-burn (RQL) combustor concept was introduced in 1980 by Mosier et al. [1] as a strategy to reduce oxides of nitrogen (NO_x) emissions from stationary gas turbine engines using fuels that were high in fuel-bound nitrogen. Later in the 1980's, studies co-sponsored by the National Aeronautics and Space Administration (NASA) and the Environmental Protection Agency (EPA), continued [2–5] that were motivated by the RQL's potential for having low NO_xemissions for these fuels. A study of the effect on NO_x emissions from fuels having high amounts of fuel-bound nitrogen was reported by Mao and Barth [6] in a staged combustor using CH_3NH_2 doping to simulate fuel-bound nitrogen.

In the 1990's the RQL concept was selected by the High Speed Research (HSR) program at NASA as a candidate for the reduction of oxides of nitrogen in engines for a High Speed Civil Transport (HSCT) [7–11] in part because in aeroengine applications an RQL may be preferred over lean premixed prevaporized (LPP) options due to safety considerations and overall performance (e.g. stability) of the RQL throughout the duty cycle. Today, RQL technology is used in aeroengines deployed commercially.

In stationary applications, lean premixed combustor technology is the standard. Safety considerations are not as severe, the duty cycle is more constrained, and the reduction in NO_x emissions is more substantial for LPP combustors in contrast to RQL technology. However, RQL combustor technology is of growing interest for stationary applications due to the attributes of more effectively processing fuels of complex and varying composition. The latter is becoming more important with the increasing international competition for fuels in general, and an increasing interest in liquefied natural gas, biomass fuels, and "opportunity fuels" (e.g., land-fill gases, digester gases, well-head gases) to either complement domestic fuel sources or serve as the sole source to a large region of a country (e.g. Southern California) or a country as a whole (e.g. South Korea).

1.1 RQL concept

The RQL (Fig. 1) is based on the premise that the primary zone of a gas turbine combustor operates most effectively with rich mixture ratios in the primary zone. A "rich-burn" condition ($\phi > 1$; see Fig. 2) in the primary zone (1) enhances the stability of the combustion reaction by producing and sustaining a high concentration of energetic hydrogen and hydrocarbon radical species, and (2) minimizes the production of nitrogen oxides due to the relatively low temperatures and low population of oxygen containing intermediate species. The effluent exiting from a rich primary zone will be high in the concentration of hydrogen (H_2), carbon monoxide (CO), and partially oxidized and partially pyrolyzed hydrocarbon species (HC). As a result, the effluent will contain both a substantial fuel value and pollutant composition, and cannot be exhausted without further processing.

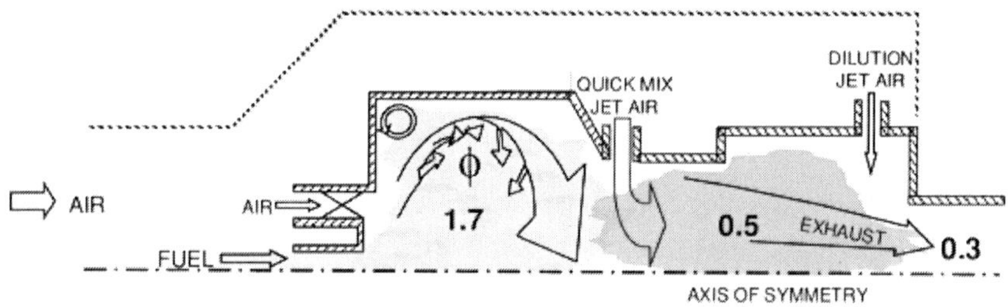

Figure 1. Rich-Burn/Quick-Mix/Lean-burn combustor ϕ, equivalence ratio)

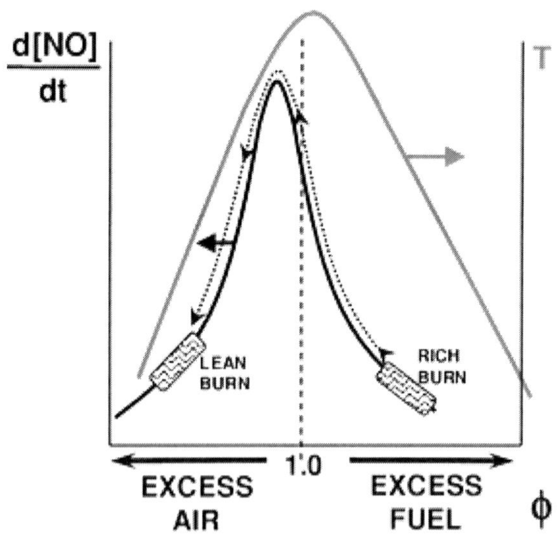

Figure 2. RQL strategy

The addition of oxygen in the mixing zone quickly converts the non-oxygen containing species in the Total Fixed Nitrogen (TFN) in the rich zone effluent to NO_x[6], but is needed to oxidize the high concentrations of carbon monoxide, hydrogen, and hydrocarbon intermediates. If no additional NO_x is formed in the mixing and lean zones, the NO_x concentration at the combustor exit would be approximately equal to the sum of the mole fractions of the prompt-NO and the major non-oxygen containing species in the rich zone effluent times the rich-to-total mass-flow ratio. Thus, the goal in achieving low NO_x in an RQL combustor is to mix air into the rich zone effluent quickly and in such a manner as to minimize the formation of thermal-NO_x to achieve an exhaust at the exit of the combustor that is comprised of the major products of combustion (CO_2, H_2O, N_2, O_2) and a near-zero concentration of pollutants (e.g., NO_x, CO, HC).

The mixing of the injected air takes the reaction through the conditions most vulnerable for the high production of oxides of nitrogen (see Fig. 2). Because the challenge is to rapidly mix air ($\phi = 0$) into the rich effluent ($\phi > 1$) to quickly create a lean composition ($\phi < 1$), the moniker "Quick-mix" is adopted to emphasize this requirement. A major candidate is a row of jets in crossflow (as shown schematically in Fig. 3) which has been found to provide rapid mixing.

Figure 3. Schmetic of laboratory model combustor

1.2 Jet mixing

To address the conditions particular to applications of jets in crossflow (JIC), numerous studies have been conducted to yield insight on jet structure and penetration, jet entrainment of crossflow fluid, and the flow field distributions resulting from jet mixing [12–20]. These studies have investigated single unbounded jets, both bounded and unbounded multiple jets, asymmetrically bounded jets, heated jets, a heated mainflow, reacting flows, or the doping of either the jets or mainflow with a tracer to allow the measurement of scalars in the flow downstream of the jet orifices. While a variety of jet orifice shapes have been studied (e.g., triangular, tear-drop, square, rectangular, bluff, slanted, and streamlined slots), no option has been identified that penetrates significantly farther or mixes faster than a round jet in an unbounded crossflow.

For a confined flow, as in a gas turbine combustor, design methods for both rectangular and cylindrical configurations have been identified at Cranfield and NASA. These are described by Lefebvre [21].

1.2.1 Non-reacting JIC mixing

Historically, an early correlation (circa 1960) for the maximum penetration of multiple jets injected normal to a crossflow was developed by Norster and is given by Eqs. (4–20) in [21]:

$$Y_{max} = 1.25\left(d_j\right)\left(J\right)^{0.5} / \left(1 + MR\right) \tag{1}$$

In an RQL mixing section, the *MR* is much higher (~10×) than in the dilution zone of a conventional combustor. Since the density ratio *DR* $(= \rho_J/\rho_M)$ and momentum-flux ratio *J* $(= (\rho_J V_J{}^2)/(\rho_M U_M{}^2))$ are about the same for the two applications, the biggest difference between the jets in conventional and RQL combustors is orifice size.

While the cylindrical geometry has been extensively researched and is directly relevant to combustor can configurations, modern large annular combustor configurations have spawned investigations of jets in crossflow in rectangular geometries.

Insight on scalar distributions resulting from mixing of confined jets in crossflow stemmed from the studies conducted in the 1970's and 1980's to address the mixing behavior associated with dilution jets in conventional gas turbine combustors. The extensive non-reacting studies of rows of jets in crossflow in rectangular, annular, and cylindrical ducts that are summarized in Holdeman [14] were funded by NASA to evaluate geometrical features (e.g., orifice spacing, opposed rows of orifices, axial staging of orifices, orifice shape, flow area convergence, effects of flow-area curvature, and a non-uniform mainstream scalar) and operating conditions (e.g. density ratio, momentum-flux ratio, mass-flow ratio) with the goal to optimize the mixing in a gas turbine combustor. Later, the NASA JIC empirical model from [14] (without the curvature effects) was implemented in the Excel® spreadsheet reported in [15–18].

For rectangular, annular, or cylindrical ducts the jet spacing *S/H* was observed to be inversely proportion to the square root of the momentum-flux ratio *J* such that:

$$\mathrm{C} = (S/H)(J)^{0.5} \tag{2}$$

For **one-side injection** optimum jet penetration was observed when C was ~2.5; significant underpenetration was observed for C less than half of the optimum value; significant overpenetration was observed when C was more than double the optimum value. For **opposed rows**, corresponding C values are ~half these (i.e. $C_{optimum}$ ~1.25; for significant underpenetration C is <0.63; for significant overpenetration C is over 2.5). For a rectangular or annular duct the jets can be injected from one side, or there can be opposed rows of aligned or staggered jets. For multiple jets in a cylindrical duct, injection is always one sided and the duct height [e.g. H in Eq. (2)] is the radius of the can.

If one assumes that the orifice spacing given by Eq. (2) should be applied to a can at the radius that divides the cylindrical cross section into equal area cylindrical and annular sections, one can derive a relation for the optimum number of holes in a cylindrical duct from Eq. (2) [14]. For C = 2.5:

$$n = \pi(2J)^{0.5}/2.5 \tag{3}$$

Traditionally, studies such as Hatch et al. [22] have defined "optimal mixing" as the shortest distance downstream from the jet orifice needed to obtain a uniform profile for key scalars (e.g., temperature, species concentration). The hypothesis is that the optimal mixing defined in this manner will minimize the production of oxides of nitrogen. Due to the complex set of variables, a design of experiments statistical approach was used by Kroll et al. [23] to select optimum variables from among the multiple factors that can affect jet mixing. In addition to non-reacting experiments and use of design of experiments methods, empirical modeling and CFD have been effectively employed both independent of and in conjunction with experiments.

The studies NASA funded in the 1990's focused on optimizing the mixing section in the Rich burn/Quick mix/Lean burn (RQL) combustor scheme in rectangular, annular, and cylindrical ducts for both non-reacting and reacting flows [19, 20]. These papers included design procedures for achieving the most rapid mixing for both cylindrical [19] and rectangular ducts [20].

1.2.2 Reacting JIC mixing

Reacting flow studies of jets in a crossflow have been conducted in both conventional and RQL combustor configurations to complement the non-reacting studies and assess the impact of heat release on the mixing processes [24–29].

For the RQL-motivated studies at atmospheric pressure by Leong et al. [28, 29], a mixture of propane and air was employed to generate a representative rich-burn effluent, and a specially designed "conditioning" section was used to create a uniform presentation (e.g., temperature, velocity, composition, and concentration) of the rich effluent to the Quick-mix section. The injection of the jet air into the rich effluent results in a combination of mixing and reaction between the rich-burn effluent and the jet air (Fig. 3). Measurements of temperature, species composition, and species concentration can then be made downstream of the jet orifices in order to establish the efficacy of mixing as a function of downstream distance.

The results from the reacting experiments reveal that distributions of a conserved scalar from non-reacting experiments generally provide a satisfactory description of the mixing of a conserved scalar for reacting jets in a crossflow. Overall, the jets in a cylindrical duct need to penetrate to approximately half of the can radius in order to maximize the mixing and avoid either under-penetration or over-penetration.

The studies by Leong et al. [28, 29] reported results for reacting flow experiments in a cylindrical duct at atmospheric pressure without inlet air preheat. A carbon balance technique similar to that of Jones et al. [30] was used by Leong et al. [28,29] to obtain the equivalence ratio. Alternate methods for obtaining the local equivalence ratio in a reacting flow were investigated by Demayo et al. [31].

The first exploration of the relation between mixing and NO_x emissions in an RQL-motivated study utilized the results of non-reacting experiments, and superimposed analytically the kinetics of NO_x production [32]. The results suggested that the best mixer might not minimize NO_x emissions.

Experimental measurements under reacting conditions at elevated pressure were provided by Meisl et al. [33]. The inlet air was preheated but the temperature was not varied, nor were the number of orifices in the mixing section varied. The study focused on the effect of varying pressure on NO_x emissions. The authors of [33] noted the two well-known sources for the emissions of NO_x, namely prompt- and thermal-NO_x. The first of these is the formation of NO in the fuel-rich primary zone [34], a contribution that Meisl et al. [33] found was relatively insensitive to pressure. The second source is the formation of thermal NO_x that may occur in the Quick-mix section which increases significantly with pressure. The study by Meisl et al. [33] also concluded that NO_x emissions are minimized by optimal mixing.

A study by Vardakas et al. [35] extended the work reported by Leong et al. [28, 29] to address the effects on both mixing and emissions of varying inlet air preheat and varying the number of orifices. This study investigated configurations with 8, 12, 14, and 22 round holes at J = 57 to give over- to optimum to under-penetrating jets. The experiments were conducted at atmospheric pressure with the jet and main air preheated independently. Selected experimental results from Vardakas et al. [35] were published in Holdeman et al. [36].

The primary conclusion based on the experimental results was that mainstream air preheat increases NO_x more that jet air preheat, and that an “optimum” mixer may not minimize NO_x emissions.

2 EXPERIMENT

2.1 Facility

The experimental facility used in this study consisted of a premixing zone, a fuel-rich combustion zone, and a jet-mixing section as shown in Fig. 4. In the premixing zone, propane gas was mixed with air upstream of the ignition point. Fuel-rich combustion was stabilized downstream of the quarl by a swirl-induced recirculation zone. To dissipate the swirl in the flow and to introduce a uniform non-swirling flow into the mixing section, the fuel-rich product was passed through an oxide-bonded silicon carbide (OBSiC) ceramic foam matrix (Hi-Tech Ceramics) with a rated porosity of 10 pores/in.

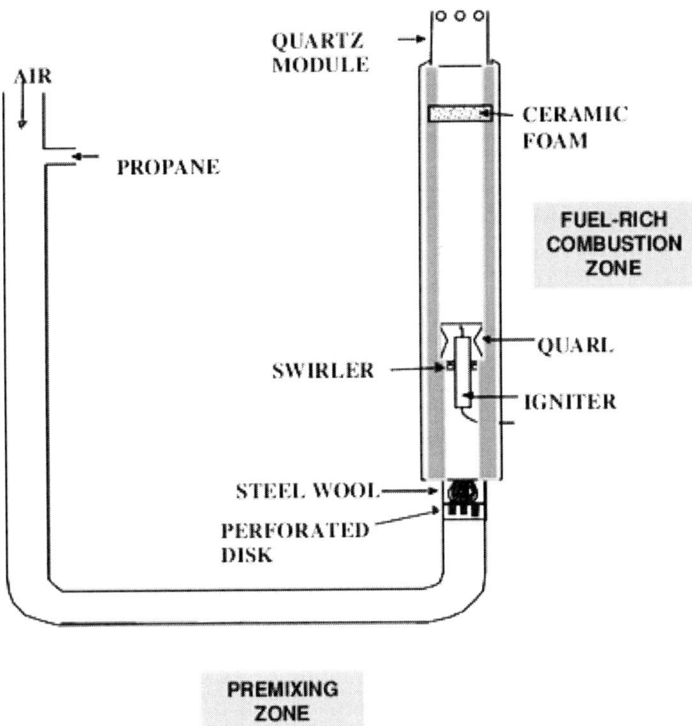

Figure 4. Schematic of experimental rich product generator with quartz RQL module

The mixing section was modular and included a cylindrical section through which the mainstream effluent passed, and to which jet air was supplied from a surrounding plenum. The plenum for the jet air was fed by four equally-spaced air ports located toward the base of the plenum. A high-temperature steel flow-straightening device installed in the plenum conditioned and equally distributed the jet air entering the mixing module.

The mixing modules were 280 mm (11 in.) in length with inner and outer diameters of 80 mm (3.15 in.) and 85 mm (3.35 in.). The row of orifices was positioned with its centerline 115 mm (4.5 in.) downstream from the module entrance. An alumina-silica blend of ceramic fiber paper provided sealing between the module and the stainless steel mating surfaces to form the plenum for the air jets.

Recirculating heaters of 20 and 25 kW were utilized to supply the necessary preheat to the main and jet air lines respectively. The preheat temperatures were independently established and controlled.

2.2 Measurements

For each module, species concentration measurements were obtained in a two-jet sector for a plane at $x/R = 1$ (plane 5 in Fig. 5a) where x/R was measured from the leading edge of the mixing module orifices. Note that $x/d = R/d$ for the data reported herein because $x = R$.

Each planar grid consisted of 16 points spread over a sector that included two orifices (Fig. 5b). The points included one point located at the center of the duct, and five points along each of the arcs at $r/R = 1/3$, $2/3$, and 1.

The measurement points along each arc were distributed such that two points were aligned with the center of the orifices and three were aligned with the midpoint between orifice centers for all cases.

Species concentration measurements were obtained by sampling through a water-cooled stainless-steel probe by routing the sample through a heated line connected to the emission analyzers. Water was condensed from the gas before the sample was analyzed by non-dispersed infrared (NDIR) analysis for CO and CO_2, paramagnetic analysis for O_2, flame ionization detection (FID) for total hydrocarbons, and chemiluminesence (CLD) for NO_x. Data are reported as measured and were not "corrected" for either the ambient humidity or the expected water content of the flow.

The extractive emissions measurement protocol was conventional. The analyzers were standardized equipment which was checked frequently with span gases. As for the relation between what exists in the flow and what is measured; this is a classical question, and they are assumed to be the same when standard procedures are followed.

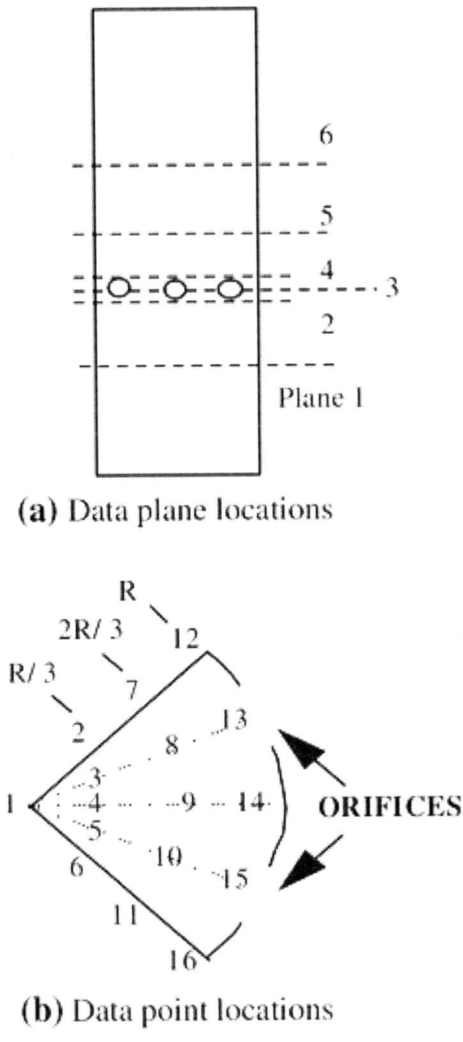

Figure 5. Measurement locations

2.3 Probe design

A double-jacked water-cooled stainless steel probe 762 mm (30 in.) in length was used to extract gas samples from the quick mixing section (Fig. 6). The probe measured 8 mm in outer diameter and tapered to 3.2 mm at the tip. A 45 degree bend was made 25.4 mm (1 in.) from the tip. The probe design was influenced by the research in [23] that reported that a thermocouple probe with a 45-degree angled tip was best as it biased the mainstream and jet flows equally in the orifice region. The probe location was fixed and the rig was traversed to obtain the measurements. The plane of the angled probe tip was positioned such that the tip was pointed toward the center of the sector wall.

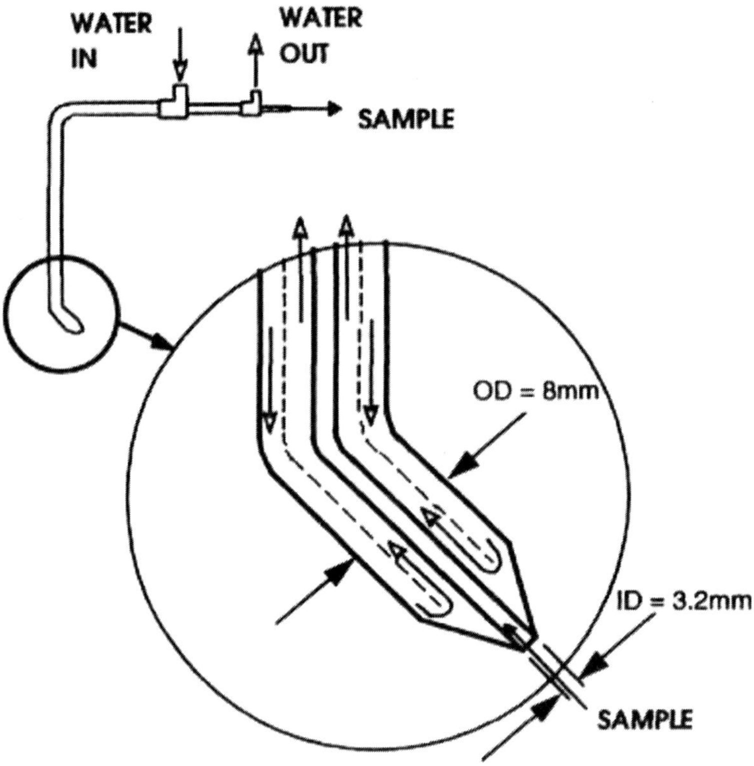

Figure 6. Probe design

2.4 Measurement uncertainty

A measure of the uncertainty of the data is given by analyzing the species measurements to determine the midplane and centerplane means and variances [36]. In a perfect world, the 3 midplane measurements would be equal and the 2 centerplane measurements would be equal at a given radius, but they often aren't. Their variability is a measure of both spatial and species uncertainty in the measurements. This is discussed further in [36].

2.5 Experimental conditions

The experiments were performed for a jet-to-mainstream momentum-flux ratio (*J*) of 57 and a mass-flow ratio (*MR*) of 2.5. The total effective area of the mixing module orifices is 9 cm^2 (1.4 $in.^2$). The ratio of the total effective jet area to the cross-sectional area of the duct is 0.18. The fuel-rich equivalence ratio and overall equivalence ratio are 1.66, and 0.45, respectively. The operating pressure for the system is one atmosphere.

The experimental results presented show comparisons between non-preheated air and preheated jet and main air cases. Inlet temperatures and other operating conditions for the experiment are noted in Table 1.

Table 1. Operating conditions

Parameter	Value
Momentum–flux ratio, J	57
Mass-flow ratio, MR	2.5
Discharge coefficient, C_d	0.72
(Total jet area)/(cross-sectional area of duct), A_J/A_M	0.25
Non-preheated inlet air temperature	22 °C (72 °F)
Inlet preheat temperature	260 °C (500 °F)

3 ANALYSIS

These analyses supplement the experimental results reported in [35, 36] with chemical equilibrium and chemical kinetics calculations to gain insight into the formation of NO_x in the rich [34], mixing, and lean (called the 2nd stage in [6]) stages.

The NASA Equilibrium Code [37] was used to perform chemical equilibrium calculations with the thermodynamic data base of Kee et al. [38]. The modeling assumptions made were for fixed equivalence ratios and constant enthalpy.

Chemical kinetic calculations were performed using the SENKIN computer code, and the CHEMKIN III Subroutines [39, 40]. The chemical kinetic mechanism of Miller and Bowman [41] was used to describe the lower hydrocarbon combustion and NO_x formation chemical kinetics along with the propane oxidation mechanism of Sundaram and Froment [42]. The conditions of the model represented those of the experiment (e.g. temperature and stoichiometry in each zone) except that the model did not account for any effects due to mixing and dispersion.

4 RESULTS AND DISCUSSION

NO_x data were obtained to investigate the effect of preheating jet and main air flows in the RQL combustor. Data for the 12 round-hole quick-mixing module are presented here as in [35, 36] for conditions without preheat and with both main and jet air preheated to the same temperature. Chemical equilibrium and kinetic modeling calculations were performed to gain insight into the NO_x formation mechanisms governing the experiments and to compare calculated levels to measured NO_x concentrations.

4.1 Chemical equilibrium

Chemical equilibrium results are presented in Fig. 7, illustrating the dominant fixed nitrogen species participating in NO_x formation, destruction, and emission mechanisms in the RQL combustor. Note that nitric oxide (NO) dominates the equilibrium species that are present for most of the equivalence ratios presented ($0.4 < \phi < 1.66$). Ammonia (NH_3) and hydrogen cyanide (HCN) are increasingly important only in the fuel-rich regime ($\phi > 1$) while nitric oxide (NO), nitrogen dioxide (NO_2), and nitrous oxide (N_2O) are the major fixed nitrogen species for fuel-lean conditions ($\phi < 1$). An important feature in these results is the existence of a minimum for total fixed nitrogen (TFN) species. The equivalence ratio corresponding to this minimum for the non-preheat air case is approximately 1.66, the operating fuel-rich zone equivalence ratio used in the experiment. As the inlet air temperature increases, however, the condition of minimum TFN shifts to higher equivalence ratios (more fuel-rich conditions) as shown in Fig. 8.

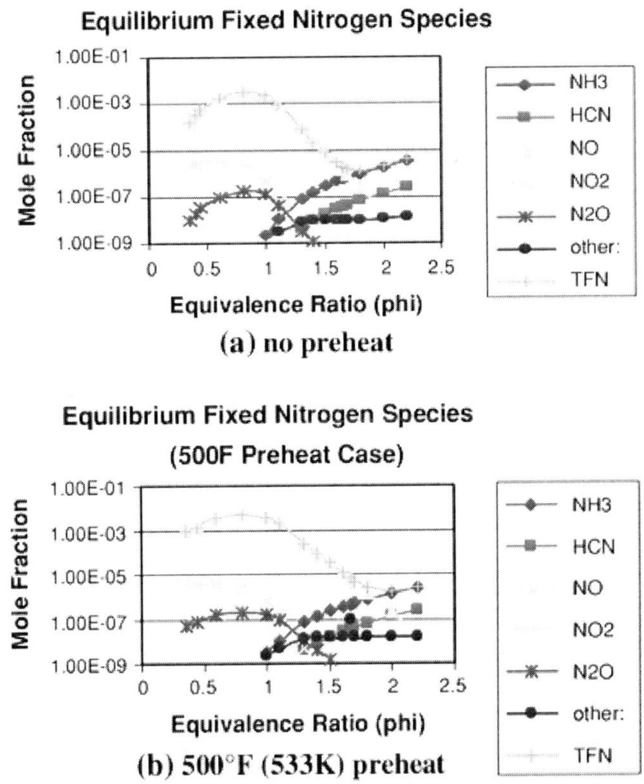

Figure 7. Chemical equilibrium fixed nitrogen species results for the non-preheat and 500 °F (533 K) preheat cases

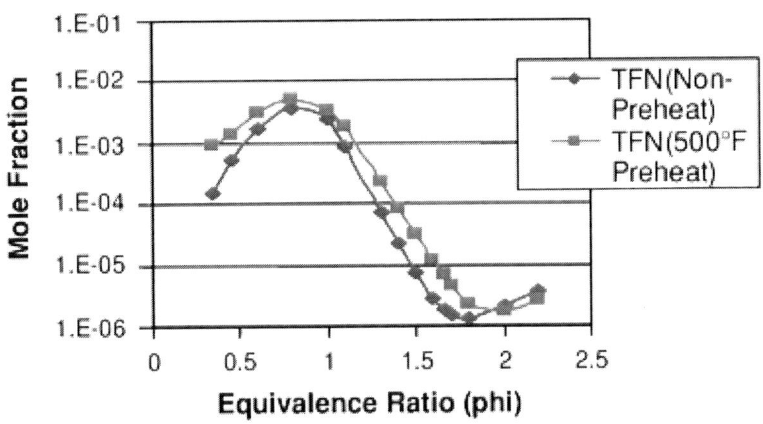

Figure 8. Total fixed nitrogen comparison between non-preheat and preheat conditions

Mao and Barth [6] reported that the minimum in TFN shifted to larger rich zone ϕ's and higher NO concentrations as the loading of CH_3NH_2 in the feed increased. As there is no fuel-bound nitrogen in the fuel used in our study, we would expect to have rich zone minimums in both the ϕ for the TFN minimum and the NO concentration there.

Figure 9 shows rich zone experimental data for the non-preheated air and 500 °F (533 K) preheated air cases. For the 500 °F case, the minimum TFN corresponds to $\phi = 2.0$ with a higher equilibrium TFN at $\phi = 1.66$ (approximately 8 ppm versus 2 ppm). The experimental NO_x data at the exit of the fuel-rich zone in the RQL combustor in Fig. 9 illustrates the effect of preheated main air on NO_x in the fuel-rich zone, and is consistent with the equilibrium results of Fig. 8. Thus, one would expect increased NO_x emissions in the preheat case simply because one was no longer operating the fuel-rich zone at the optimal stoichiometry to minimize TFN for preheated air conditions.

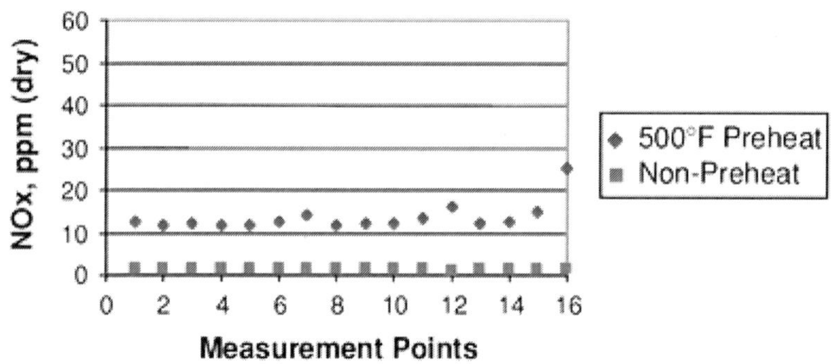

Figure 9 NO_x Measurements in RQL fuel-rich zone (x/R = −1) for preheat and non-preheat conditions

4.2 Chemical kinetics

The equivalence ratios used for the chemical kinetic modeling correspond to operating equivalence ratios for the fuel-rich ($\phi > 1$) and fuel-lean ($\phi <$!) zones of the RQL combustor. Chemical kinetic calculations were also performed for stoichiometric ($\phi = 1$) conditions because of their importance to NO_x formation in transitioning from fuel-rich to fuel-lean zones of the combustor which occurs in the quick-mix zone. The adiabatic flame temperature was used as a baseline case for each of the chemical kinetic simulations. In addition, two lower temperatures were simulated to account for various levels of heat loss that might occur in practical systems.

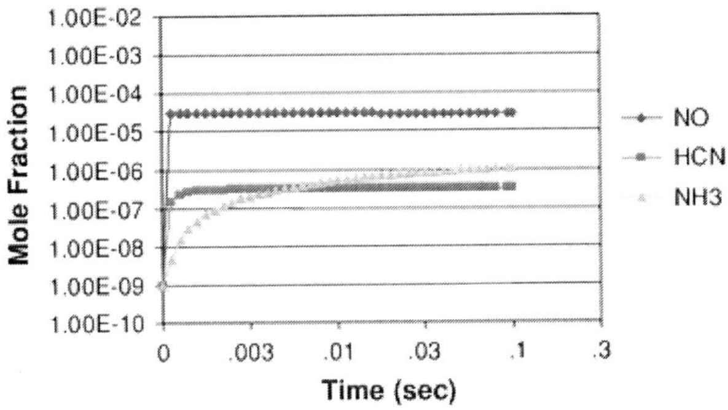

(a) Temperature = 2000 K, ϕ = 1.66

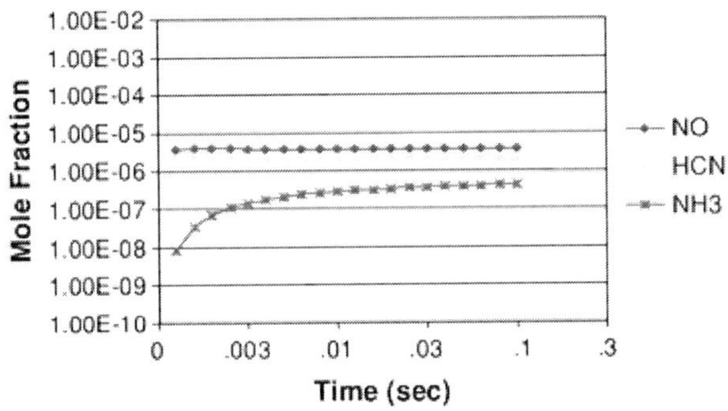

(b) Temperature = 1573K, ϕ = 1.66

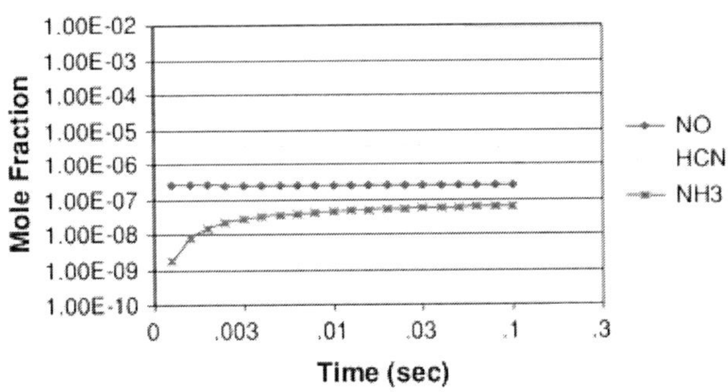

(c) Temperature = 1425 K, ϕ = 1.66

Figure 10. Chemical kinetic evolution of primary fixed nitrogen species for ϕ = 1.66 conditions similar to RQL fuel-rich zone

Measurements of temperature that were made in the RQL combustor were consistently bounded by the chemical kinetic modeling conditions.

4.2.1 $\phi = 1.66$ (Fuel-rich zone results)

The chemical kinetic results for the fuel-rich conditions of $\phi = 1.66$ are presented in Fig. 10 for constant temperature conditions of 2,000, 1,573, and 1,425 K. The major fixed nitrogen species identified by the model are nitric oxide (NO), hydrogen cyanide (HCN) and ammonia (NH_3) which is consistent with the chemical equilibrium results of Fig. 4. In all cases (adiabatic or with heat-loss) the NO and HCN kinetics are quite rapid leading to near equilibrium concentrations in less than 2 ms. The ammonia concentration continually increases with time and temperature in the range investigated.

4.2.2 $\phi = 1.00$ (Stoichiometric results)

As air is introduced to the fuel-rich zone effluent in the quick-mix section of the RQL combustor there is a transition in overall equivalence ratio from rich ($\phi > 1$) to lean ($\phi < 1$). This transition does not occur instantaneously, but rather is characterized by finite mixing times resulting in spatial and temporal inhomogeneities and fluctuations between these two stoichiometries. During this mixing process stoichiometric or near-stoichiometric conditions and high temperatures can exist locally with a corresponding increase in potential for the formation of NO_x and other fixed nitrogen species.

Figure 11 shows chemical kinetics calculations for stoichiometric conditions. As in the previous case adiabatic conditions are shown (2,364 K here) and the other two temperatures represent levels of heat loss which bound temperatures measured in the experiment in the quick-mix zone. Note that the adiabatic temperature for stoichiometric conditions is higher than that for rich conditions.

In Fig. 11, the rate of NO production under stoichiometric conditions is shown to be exceedingly fast. The other fixed nitrogen species which are significant for these conditions include nitrous oxide (N_2O) and nitrogen dioxide (NO_2), both with significantly lower concentrations compared to NO. NO forms very rapidly for all temperature conditions investigated. As with the rich conditions in Fig. 10, NO concentration levels for stoichiometric conditions shown in Fig. 11 are very dependent on temperature with mole fractions of over 1E – 03 (1,000 ppm) for the 2,364 K case, less than 1E – 03 (1,000 ppm) for the 2,215 K case, and approximately 1E – 04 (100 ppm) for the 2,103 K case, all obtained within 2 ms. The N_2O and NO_2 species are produced almost as quickly as the NO, but stay consistently below a mole fraction of 1E – 06 (1 ppm). The NO results show that the concentration level in stoichiometric regions can range from a mole fraction of 1E – 03 (1,000 ppm) to 1E – 04 (100 ppm) depending on temperature.

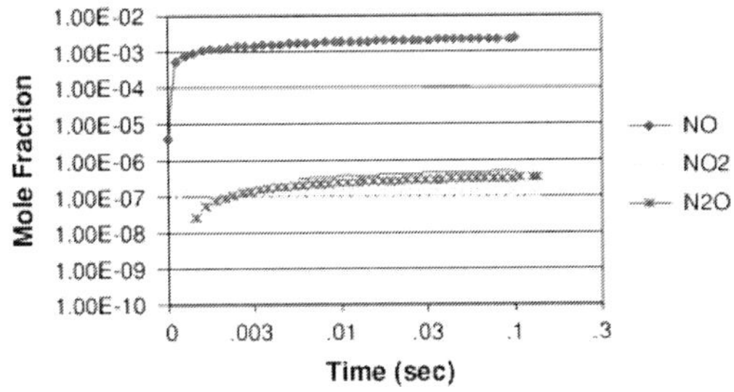

(a) Temperature = 2364 K, ϕ = 1.00

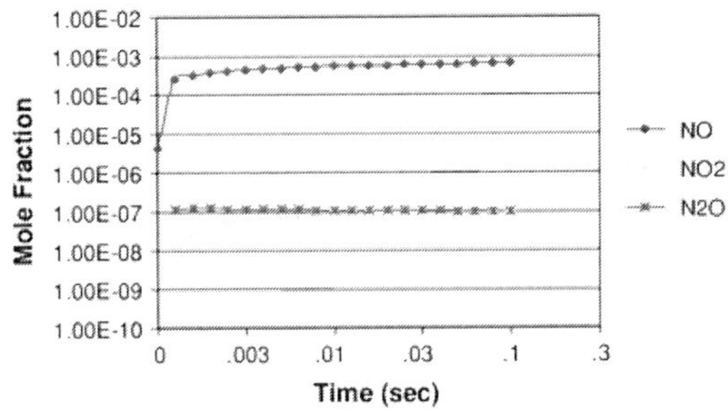

(b) Temperature = 2215 K, ϕ = 1.00

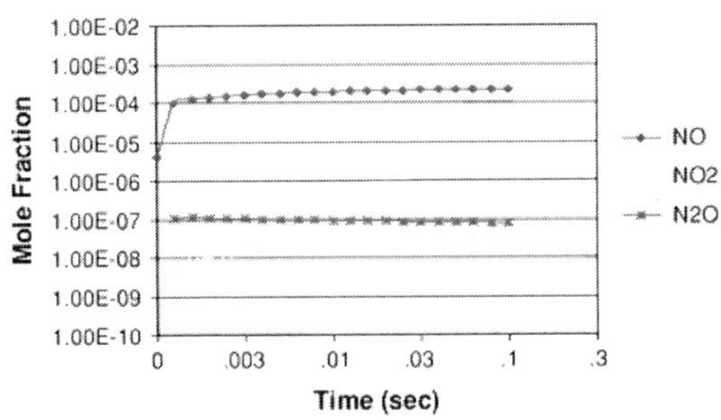

(c) Temperature = 2103 K, ϕ = 1.00

Figure 11. Chemical kinetic evolution of primary fixed nitrogen species for ϕ = 1.00

Although the thermal-NO_x mechanism can produce substantial NO_x in the quick-mix zone, the objective is to minimize it so the combustor NO_x emissions are dominated by contributions from the prompt-NO_x and fuel-NO_x mechanisms.

4.2.3 ϕ = 0.45 (Fuel-lean zone results)

Chemical kinetic calculations representing the fuel-lean zone of the RQL combustor are presented in Fig. 12. The simulation was performed by addition of sufficient air to the products of the fuel-rich zone (ϕ = 1.66) to create the ϕ = 0.45 conditions. Although the formation of NO is rapid for these conditions, the NO concentrations that are observed are significantly lower than those observed for either the stoichiometric or fuel-rich conditions. This is due to the lower temperatures of the fuel-lean zone (the excess air dilutes the mixture and lowers the temperature). The results of Fig. 12 indicate that most of the NO_x is not formed in the fully mixed fuel-lean zone of the combustor since single digit levels are all that are produced for ϕ = 0.45 even with a long residence time.

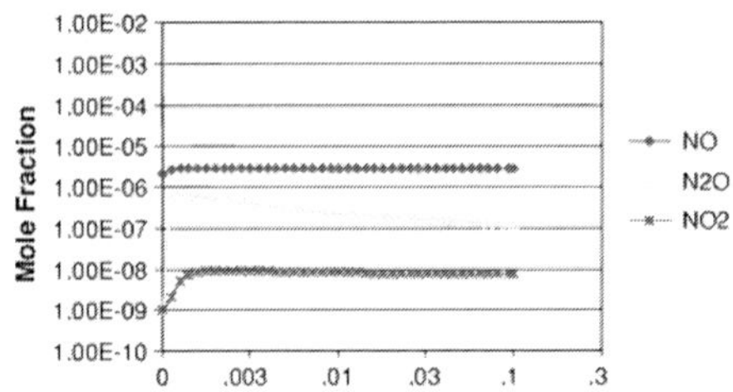

(a) Temperature = 1600K, ϕ = 0.45

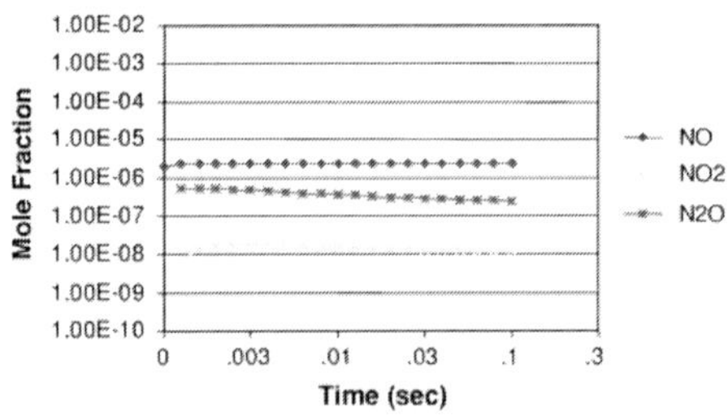

(b) Temperature = 1450 K, ϕ = 0.45

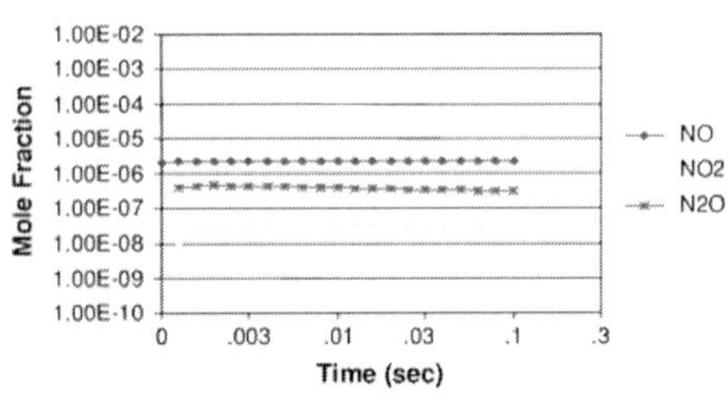

(c) Temp = 1350K, ϕ = 0.45

Figure 12. Chemical kinetic evolution of primary fixed nitrogen species for ϕ = 0.45. Conditions similar to conditions in the fuel-lean zone

4.3 Experimental results

Experimental data from Vardakas et al. [35] are shown in parts (a) and (b) of Fig. 13for the non-preheat and main & jet preheat cases using the 12-hole module. The data shown in Fig. 13 indicate a significant increase in NO_x for the preheated case. A comparison of the various modules (8-, 12-, 14-, and 22-round-hole) for all preheat conditions is presented in Vardakas et al. [35].

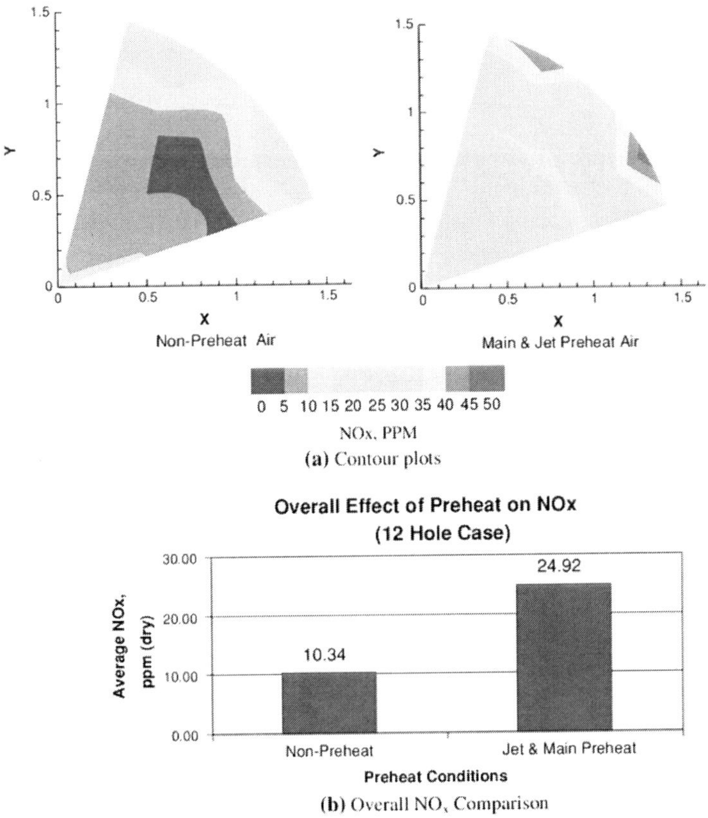

Figure 13. Experimental NO_x data at x/R = 1 and J = 57 for the 12 hole module for non-preheat and jet and main air preheat conditions

Figure 13(a) presents contour plots of the NO_x distribution for conditions without preheat and with both streams preheated. These plots show the effect that jet penetration has on determining the locations of peak NO_x concentration. The contour plots indicate that concentrations are high in the shear layers on both sides on the maximum jet penetration but the highest NO_x concentrations occur near the wall of the module in the wakes of the jets.

The effect of main & jet air preheat on overall NO_x emissions is presented in the bar chart in Fig. 13b and shows a large increase in overall NO_x emissions with both streams preheated. The results of Vardakas et al. [35] showed a significantly larger increase in NO_x emissions with both streams preheated than there was with no preheat and when only the jet air was preheated.

The variations of the averages for the cases shown vary from a low of ~10 ppm for the case without preheat to a high of ~25 ppm with both streams preheated.

NO_x can be produced by several mechanisms (prompt-, fuel-, and thermal-NO_xmechanisms), but when thermal NO_x formation is minimized, the NO_x is being produced primarily in the fuel-rich zone, and through conversion from other TFN species in the early mixing stage. The fuel-NO_x mechanism can convert the fuel-rich zone total fixed nitrogen (TFN) species other than NO (e.g., NH_3, HCN), which are the primary non-oxygen containing nitrogen species in the fuel-rich zone, to $NO_x$with relatively high conversion [6]. Although the chemical modeling does not take into account the effects of the mixing jets, the modeling results suggest that most of the NO_x observed in the experimental data is formed either in the quick-mix zone or in the fuel-rich zone.

This information can be used to devise technique(s) to further reduce overall NO_xformation in the RQL combustor. With preheat NO_x formation in the fuel-rich zone is higher than optimal because the optimal stoichiometry for minimum TFN shifts to more fuel-rich conditions as temperatures increase. Temperature based control of the fuel-rich zone stoichiometry is suggested.

The increased dependence of overall NO_x emissions on main air preheat suggests another optimization of the RQL concept. NO_x could be reduced by increasing heat transfer from the fuel-rich zone to the fuel-lean zone. One strategy could be to pass the jet air through the fuel-rich zone combustor liner to transfer heat from this zone to the quick-mix and fuel-lean zones which do not have as significant a dependence upon preheat temperature.

The distributions of equivalence ratios at $x/R = 1$ and $J = 57$ for the 12-hole module are given in Fig. 14. Preheat has almost no effect on jet penetration and mixing as expected, and the equivalence ratio distributions are quite uniform since the 12 hole configuration is a nearly optimum mixer at $J = 57$. As in Leong et al. [28, 29] the distributions in these plots were determined by noting that carbon atoms in carbon containing species are conserved [30].

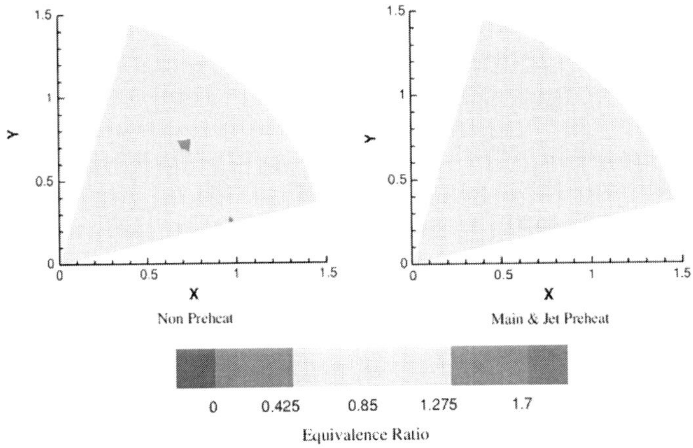

Figure 14. Jet Equivalence Ratio Distribution Plots at x/R = 1 and J = 57 for the 12 round-hole module for non-preheat and jet & main air preheat conditions. The fuel-rich equivalence ratio and overall equivalence ratio are 1.66, and 0.45, respectively

Note that the completely mixed condition for a conserved scalar at $MR = 2.5$ would be $\phi = 0.45$. Unmixed mainstream fluid would be $\phi = 1.66$ and unmixed jet fluid would be $\phi = 0$.

5 SUMMARY AND CONCLUSIONS

An experiment was performed to expand upon earlier work by incorporating preheated inlet air, and examining the effect it has on NO_x production. To further understand the mechanisms involved, chemical equilibrium and kinetic modeling was done. The following was revealed in this study:

- An important feature in these results is the existence of a minimum for total fixed nitrogen (TFN) species.
- Nitric oxide (NO) dominates the TFN species that are present for most equivalence ratios from $\phi = 0.4$ to $\phi = 1.66$.
- Ammonia (NH_3) and hydrogen cyanide (HCN) are increasingly important only in the fuel-rich regime.
- Under air preheat conditions, the fuel-rich zone should be operated at a higher equivalence ratio to reduce TFN production in this zone.

- NO_x is formed in less than 2 ms at all conditions.
- There exists sufficient residence time for significant amounts of NO to be formed in each zone.
- An optimization of the RQL concept could include increases in the heat transfer from the fuel-rich zone to the mixing and fuel-lean zones and temperature-based control of fuel-rich zone stoichiometry.

These conclusions are based upon chemical equilibrium, chemical kinetics, and experimental results that point to primary production of NO_x in the fuel-rich and early mixing zones. The mechanisms that lead to this NO_x production and emissions are prompt-NO_x and fuel-NO_x. Thermal-NO_x formed in the mixing zone can dominate emissions so NO_x formation via this mechanism must be minimized in an RQL combustor.

ACKNOWLEDGMENTS

The experimental work was supported by cooperation agreement NCC 3-412 from the NASA Glenn Research Center and the complementary chemical equilibrium and kinetics calculations were funded by the UCI Combustion Laboratory (UCICL) at the University of California, Irvine. The authors would also like to thank the Combustion Branch at the NASA Glenn Research Center for providing for the Open Access publication and color printing of this paper.

REFERENCES

1. Mosier SA, Pierce RM (1980) Advanced combustor systems for stationary gas turbine engines, phase I. Review and preliminary evaluation, vol I, contract 68-02-2136, FR-11405, Final Report, U.S. Environmental Protection Agency, March
2. Novick AS, Troth DL (1981) Low NO_X heavy fuel combustor concept program, NASA CR-165367 (same as DOE/NASA/0148-1), October
3. Lew HG, Carl DB, Vermes, G, DeZubay EA, Schwab JA, Prothroe D (1981) Low NO_X heavy fuel combustor concept program–phase 1: combustion technology generation, NASA CR-165482 (same as DOE/NASA/0146-1), October

4. Beal G, Hinton B, Kemp F, Russel P, Sederquist R (1981) Low NO_X heavy fuel combustor concept program, NASA CR-165512 (same as DOE/NASA/0149-3), October

5. Kemp FS, Sederquist RA, Rosfjord TJ (1982) Low NO_X heavy fuel combustor concept program–phase 1A: coal gas addendum final report, NASA CR-165577 (same as DOE/NASA/0149-) January

6. Mao F, Barat RB (1996) Minimization of NO during staged combustion of CH_3NH_2. Combust Flame 105(4):557–568

7. Shaw RJ, Gilkey S, Himes R (1993) Engine technology challenges for a 21st century high-speed civil transport, NASA TM-106216, September

8. Rosfjord TJ, Pagent FC (2001) Experimental assessment of the emissions control potential of a rich/quench/lean combustor for high speed civil transport aircraft engines, NASA CR—2001-210613, January

9. Holdeman JD, Chang CT, Nguyen HL (2001) Low Emissions RQL Flametube combustor component test results, NASA/TM—2001-210678, January

10. Chang CT, Holdeman JD, Nguyen HL (2001) Low Emissions RQL flametube combustor test results, NASA/TM–2001-211107, July

11. Peterson CO, Sowa WA, Samuelsen GS (2002) Performance of a model rich burn-quick mix-lean burn combustor at elevated temperature and pressure, NASA/CR-2002-211992, December

12. Margason RJ (1993) Fifty years of jet in cross flow research, presented at computational and experimental assessment of jets in cross flow, AGARD Conf Proc 534, April

13. Demuren AO (1994) Modeling jets in cross flow. NASA contractor report 194965, August (same as ICASE Report No. 94-71, August 1994)

14. Holdeman JD (1993) Mixing of multiple jets with a confined subsonic crossflow. Prog Energy Combust Sci 19:31–70. doi:10.1016/0360-1285(93)90021-6 (similar to AIAA Paper 91-2458 and NASA TM 104412, June 1991)

15. Holdeman JD, Smith TD, Clisset JR, Lear WE (2005) A spreadsheet for the mixing of a row of jets with a confined crossflow. NASA/TM—2005-213137, February

16. Holdeman JD, Clisset JR, Moder JP, Lear WE (2006) On the mixing of single and opposed rows of jets with a confined crossflow. NASA/TM—2006-214226, October

17. Holdeman JD, Clisset JR, Moder JP (2010) Spreadsheet calculations for jets in crossflow: opposed rows of inline and staggered holes and single and opposed rows with alternating hole sizes. NASA/TM—2010-216100, July (an excerpt was published in Heat and Mass Transfer, 48(2):412–424 (Feb 2012), doi: 10.1007/s00231-011-0913-6 and another excerpt was published in the J Eng Gas Turbines Power 132(June 2010):64502-1-7

18. Holdeman JD, Clisset JR, Moder JP (2011) Spreadsheet calculations for jets in crossflow: opposed rows of slots slanted at 45°. NASA/TM—2011-215980, December

19. Holdeman JD, Liscinsky DS, Oechsle VL, Samuelsen GS, Smith CE (1997) Mixing of multiple jets with a confined subsonic crossflow: part I—cylindrical ducts. J Eng Gas Turbines Power 119(10):852–862. doi:10.1115/1.2817065 (same as ASME Paper 96-GT-482 and NASA TM 107185, June 1996)

20. Holdeman JD, Liscinsky DS, Bain DB (1999) Mixing of multiple jets with a confined crossflow: part II–opposed rows of orifices in rectangular ducts. J Eng Gas Turbines Power 121(7):551–562. doi:10.1115/1.2818508 (same as ASME Paper 97-GT-439 and NASA TM 107461, June 1997)

21. Lefebvre AH (1999) Gas turbine combustion, 2nd edn. Taylor and Francis, New York

22. Hatch MS, Sowa WA, Samuelsen GS, Holdeman JD (1995) Jet mixing into a heated cross flow in a cylindrical duct: influence of geometry and flow variations. J Propuls and Power 11(3):393–400 (same as AIAA-92-0773 and NASA TM 105390, Jan. 1992; derived from NASA CR-194473, June 1996

23. Kroll JT, Sowa WA, Samuelsen GS, Holdeman JD (2000) Optimization of orifice geometry for cross flow mixing in a cylindrical duct. J Propuls Power 16(6):929–936 (same as AIAA-94-0219 and NASA TM 106436, Jan. 1994; derived from NASA CR-198482, April 1996)

24. Noyce JR, Sheppard CGW, Yamba FD (1981) Measurements of mixing and species concentration within a gas turbine type combustor. Combust Sci Technol 25:209–217

25. Heitor MV, Whitelaw JH (1986) Velocity, temperature, and species characteristics of the flow in a gas-turbine combustor. Combust Flame 64:1–32

26. Richards CD, Samuelsen GS (1990) The role of primary jet injection on mixing in gas turbine combustors. Twenty-Third Symp (Int) Combust 23:1071–1077

27. Bicen AF, Tse DGN, Whitelaw JH (1990) Combustion characteristics of a model can-type combustors. Combust Flame 80:111–125

28. Leong MY, Samuelsen GS, Holdeman JD (1999) Mixing of jet air with a fuel-rich, reacting crossflow. J Propuls Power 15(5):617–622 (same as WSS/CI Paper No. 97S-034 and NASA TM 107430, Apr. 1997; derived from NASA CR-195375, Sept. 1996)

29. Leong MY, Samuelsen GS, Holdeman JD (2000) Optimization of jet mixing into a rich, reacting crossflow. J Propuls Power 16(5):729–735 (also AIAA-98-0156 and NASA/TM–97-206294, Dec. 1997; see also NASA CR-195375, Sept. 1996)

30. Jones WP, McDonell V, McGuirk JJ, Milosavljevic VD, Taylor AMKP, Whitelaw JH (1993) The calculation of mean mixture fractions in turbulent non-premixed methane flames from aspiration-probe measurements. Department of Mechanical Engineering, Imperial College of Science, Technology, and Medicine, Report TFS/93/13, March

31. Demayo TN, Leong MY, Samuelsen GS, Holdeman, JD (2003) Assessing jet-induced spatial mixing in a rich, reacting crossflow. J Propuls Power 19(1):14–21 (same as NASA/CR—2004-212886, Jan. 2004)

32. Hatch MS, Sowa WA, Samuelsen GS, Holdeman JD (1992) Influence of geometry and flow variation on the NO formation in the quick mixer of a staged combustor. Accepted for publication in the J Eng Gas Turbines Power but distribution was initially restricted by HSR (same as NASA TM–105639, December 1995; derived from NASA CR-194473, June 1996)

33. Meisl J, Koch R, Kneer R, Wittig S (1994) Study of NO_x emission characteristics in pressurized staged combustor concepts. Twenty-Fifth Sym (Int) Combust 25:1043–1049

34. Fenimore CP (1976) Reactions of fuel-nitrogen in rich flame gases. Combust Flame 26:249–256

35. Vardakas MA, Leong MY, Brouwer J, Samuelsen GS, Holdeman JD (1999) The effect of air preheat at atmospheric pressure on the

formation of NO_x in the quick-mix sections of an axially staged combustor. NASA TM-1999-209431, August

36. Holdeman JD, Vardakas MA, Chang CT (2008) Mixing of multiple jets with a confined subsonic crossflow: part III—the effects of air preheat and number of orifices on flow and emissions in an RQL mixing section. NASA/TM–2008-215151, March (similar to J Fluids Eng 129(11):1460–1467)

37. Gordon S, McBride BJ (1976) Computer program for calculation of complex chemical equilibrium composition, rocket performance, incident and reflected shocks, and Chapman-Jouquet Detonations. NASA SP-273, Interim Revision

38. Kee RJ, Rupley FM, Miller JA (1990) The chemkin thermodynamic data base. Sandia National Laboratories report, SAND87-8215, Sandia Natioinal Laboratories, Albuquerque, NM

39. Lutz AE, Kee RJ, Miller JA (1987) SENKIN: a FORTRAN program for predicting homogeneous gas phase chemical kinetics with sensitivity analysis, Sandia report SAND87-8248-UC401, Combustion research facility, Sandia Natioinal Laboratories, Albuquerque, NM, July

40. Kee RJ, Rupley FM, Meeks E, Miller JA (1996) CHEMLIN III: a FORTRAN chemical kinetics package for the analysis of gas phase chemical and plasma kinetics, Sandia report, SAND96-8216-UC405. Sandia National Laboratories, Albuquerque, NM

41. Miller JA, Bowman CT (1989) Mechanism and modeling of nitrogen chemistry in combustion. Prog Energy Combust Sci 15:287–338

42. Sundaram KM, Froment GF (1979) Kinetic of coke deposition in the thermal cracking of propane. Chem Eng Sci 34:635

Index

T

Z